T0309718

Introduction to Aerospace Engineering

Introduction to Aerospace Engineering

Basic Principles of Flight

Ethirajan Rathakrishnan
Indian Institute of Technology, Kanpur

This edition first published 2021
© 2021 John Wiley & Sons, Inc.

The right of Ethirajan Rathakrishnan to be identified as the author of this work has been asserted in accordance with law.

Registered Office
John Wiley & Sons, Inc., 111 River Street, Hoboken, NJ 07030, USA

Editorial Office
111 River Street, Hoboken, NJ 07030, USA

For details of our global editorial offices, customer services, and more information about Wiley products visit us at www.wiley.com.

Wiley also publishes its books in a variety of electronic formats and by print-on-demand. Some content that appears in standard print versions of this book may not be available in other formats.

Limit of Liability/Disclaimer of Warranty

While the publisher and authors have used their best efforts in preparing this work, they make no representations or warranties with respect to the accuracy or completeness of the contents of this work and specifically disclaim all warranties, including without limitation any implied warranties of merchantability or fitness for a particular purpose. No warranty may be created or extended by sales representatives, written sales materials or promotional statements for this work. The fact that an organization, website, or product is referred to in this work as a citation and/or potential source of further information does not mean that the publisher and authors endorse the information or services the organization, website, or product may provide or recommendations it may make. This work is sold with the understanding that the publisher is not engaged in rendering professional services. The advice and strategies contained herein may not be suitable for your situation. You should consult with a specialist where appropriate. Further, readers should be aware that websites listed in this work may have changed or disappeared between when this work was written and when it is read. Neither the publisher nor authors shall be liable for any loss of profit or any other commercial damages, including but not limited to special, incidental, consequential, or other damages.

Library of Congress Cataloging-in-Publication Data

Names: Rathakrishnan, Ethirajan, author.
Title: Introduction to aerospace engineering : basic principles of flight / Ethirajan Rathakrishnan.
Description: First edition. | Hoboken, NJ : Wiley, 2022. | Includes bibliographical references and index.
Identifiers: LCCN 2021003268 (print) | LCCN 2021003269 (ebook) | ISBN 9781119807155 (cloth) | ISBN 9781119806844 (adobe pdf) | ISBN 9781119806868 (epub)
Subjects: LCSH: Aerodynamics. | Aerospace engineering.
Classification: LCC TL570 . R3295 2022 (print) | LCC TL570 (ebook) | DDC 629.132/3–dc23
LC record available at https://lccn.loc.gov/2021003268
LC ebook record available at https://lccn.loc.gov/2021003269

Cover Design: Wiley
Cover Image: © guvendemir/iStock/Getty Images

Set in 9.5/12.5pt STIXTwoText by Straive, Chennai, India

SKY10026982_051221

This book is dedicated to my parents,
Mr. Thammanur Shunmugam Ethirajan
and
Mrs. Aandaal Ethirajan

Ethirajan Rathakrishnan

Contents

Preface

This book has been developed to introduce the subject of Aerospace Engineering to the beginners. Introduction to aerospace engineering is a compulsory course for Aerospace Engineering students. This book, being the manuscript developed using the course material used in teaching this course for a long period, precisely presents the basics of theoretical and application aspects of the subject.

This book is developed based on the class tested material for the course Introduction to Aerospace Engineering, at BS and MS levels, taught by the author at Indian Institute of Technology Kanpur. The topics covered are; Basics, International Standard Atmosphere, Aircraft Configurations, Low-Speed Aerofoils, High-Lift Devices, Thrust, Level Flight, Gliding, Performance, Stability and Control, Manoeuvres, Rockets. All these topics are introduced in such a manner that the students studying these for the first time could comfortably follow and assimilate the material covered.

The material covered in this book is so designed that any beginner can follow it comfortably. The book is organised in a logical manner and the topics are discussed in a systematic manner.

My sincere thanks to my undergraduate and graduate students at Indian Institute of Technology Kanpur, who are directly and indirectly responsible for the development of this book.

My special thanks to Dr. S.M. Aravindh Kumar, Department of Aerospace Engineering, S.R.M. Institute of Science and Technology, Chennai, for critically checking the manuscript and giving some useful suggestions.

For instructors only, a companion Solutions Manual is available from John Wiley that contains typed solutions to the end-of-chapter problems.

Chennai, India
March 23, 2021

Ethirajan Rathakrishnan

About the Author

Ethirajan Rathakrishnan is professor of Aerospace Engineering at the Indian Institute of Technology Kanpur, India. He is well known internationally for his research in the area of high-speed jets. The limit for the passive control of jets, called the *Rathakrishnan Limit*, is his contribution to the field of jet research, and the concept of *breathing blunt nose (BBN)*, which simultaneously reduces the positive pressure at the nose and increases the low pressure at the base is his contribution to drag reduction at hypersonic speeds. Positioning the twin-vortex Reynolds number at around 5000, by changing the geometry from cylinder, for which the maximum limit for the Reynolds number for positioning the twin-vortex was found to be around 160, by von Karman, to flat plate, is his addition to vortex flow theory. He has published a large number of research articles in many reputed international journals. He is a Fellow of many professional societies including the Royal Aeronautical Society. Rathakrishnan serves as the Editor-in-Chief of the *International Review of Aerospace Engineering* (IREASE) and *International Review of Mechanical Engineering* (IREME) journals. He has authored 13 other books: *Gas Dynamics*, 7th ed. (PHI Learning, New Delhi, 2020); *Fundamentals of Engineering Thermodynamics*, 2nd ed. (PHI Learning, New Delhi, 2005); *Fluid Mechanics: An Introduction*, 4th ed. (PHI Learning, New Delhi, 2021); *Gas Tables*, 3rd ed. (Universities Press, Hyderabad, India, 2012); *Theory of Compressible Flows* (Maruzen Co., Ltd. Tokyo, Japan, 2008); *Gas Dynamics Work Book*, 2nd ed. (Praise Worthy Prize, Naples, Italy, 2013); *Elements of Heat Transfer* (CRC Press, Taylor & Francis Group, Boca Raton, FL, USA, 2012); *Theoretical Aerodynamics* (John Wiley, NJ, USA, 2013); *High Enthalpy Gas Dynamics* (John Wiley & Sons Inc., 2015); *Dynamique Des Gaz* (Praise Worthy Prize, Naples, Italy, 2015); and *Instrumentation, Measurements and Experiments in Fluids*, 2nd ed. (CRC Press, Taylor & Francis Group, Boca Raton, FL, USA, 2017), *Helicopter Aerodynamics* (PHI Learning, New Delhi, 2019); *Applied Gas Dynamics* 2nd ed. (John Wiley & Sons Inc., 2019).

About the Companion Website

This book is accompanied by a companion website:

www.wiley.com/go/Rathakrishnan/IntroductiontoAerospaceEngineering

The website has solutions manual and lecture slides.

1

Basics

1.1 Introduction

Aerodynamics is the study of forces and the resulting motion of objects through the air. This word is coined with the two Greek words: *aerios*, concerning the air, and *dynamis*, meaning force. Judging from the story of Daedalus and Icarus,[1] humans have been interested in aerodynamics and flying for thousands of years, although flying in a heavier-than-air machine has been possible only in the last century. Aerodynamics affects the motion of high-speed flying machines, such as aircraft and rockets, and low-speed machines, such as cars, trains, and so on. Therefore, *aerodynamics* may be described as a branch of dynamics concerned with studying the motion of air, particularly when it interacts with a solid object. Aerodynamics is a subfield of fluid dynamics and gas dynamics. It is often used synonymously with gas dynamics, with the difference being that gas dynamics applies to all gases.

Understanding the flow field around an object is essential for calculating the forces and moments acting on the object. Typical properties calculated for a flow field include velocity, pressure, density, and temperature as a function of spatial position and time. Aerodynamics allows the definition and solution of equations for the conservation of mass, momentum, and energy in air. The use of aerodynamics through mathematical analysis, empirical approximations, wind tunnel experimentation, and computer simulations forms the scientific basis for heavier-than-air flight and a number of other technologies.

Aerodynamic problems can be classified according to the flow environment. *External aerodynamics* is the study of flow around solid objects of various shapes. Evaluating the lift and drag on an airplane or the shock waves that form in front of the nose of a rocket are examples of external aerodynamics. *Internal aerodynamics* is the study of flow through passages in solid objects. For instance, internal aerodynamics encompasses the study of the airflow through a jet engine.

Aerodynamic problems can also be classified according to whether the flow speed is below, near or above the speed of sound. A problem is called *subsonic* if all the speeds in the problem are less than the speed of sound, *transonic* if speeds both below and above the speed of sound are present, *supersonic* if the flow speed is greater than the speed of sound, and *hypersonic* if the flow speed is more than five times the speed of sound.

---

1 Daedalus – his name means 'skilled worker' – was a famous architect, inventor, and master craftsman known for having created many objects that figure prominently in various myths. He had a son named Icarus. Among the many inventions and creations crafted by Daedalus were the wooden cow he constructed for the queen Pasiphae, the Labyrinth of the Minotaur at Knossos on the island of Crete, and artificial wings for himself and his son Icarus, and he was even said to have invented images.

Introduction to Aerospace Engineering: Basic Principles of Flight, First Edition. Ethirajan Rathakrishnan.
© 2021 John Wiley & Sons, Inc. Published 2021 by John Wiley & Sons, Inc.
Companion Website: www.wiley.com/go/Rathakrishnan/IntroductiontoAerospaceEngineering

The influence of viscosity in the flow dictates a third classification. Some problems may encounter only very small viscous effects on the solution; therefore the viscosity can be considered to be negligible. The approximations made in solving these problems is the viscous effect that can be regarded as negligible. These are called *inviscid flows*. Flows for which viscosity cannot be neglected are called *viscous flows*.

1.2 Overview

Humans have been harnessing aerodynamic forces for thousands of years with sailboats and windmills [1]. Images and stories of flight have appeared throughout recorded history [2], such as the legendary story of Icarus and Daedalus [3]. Although observations of some aerodynamic effects such as wind resistance (for example, drag) were recorded by Aristotle, Leonardo da Vinci, and Galileo Galilei, very little effort was made to develop a rigorous quantitative theory of airflow prior to the seventeenth century.

In 1505, Leonardo da Vinci wrote the Codex (an ancient manuscript text in book form) on the *Flight of Birds*, one of the earliest treatises on aerodynamics. He was the first to note that the centre of gravity of a flying bird does not coincide with its centre of pressure, and he describes the construction of an ornithopter with flapping wings similar to birds.

Sir Isaac Newton was the first to develop a theory of air resistance [4], making him one of the first aerodynamicists. As a part of that theory, Newton considered that drag was due to the dimensions of the body, the density of the fluid, and the velocity raised to the second power. These all turned out to be correct for low-speed flow. Newton also developed a law for the drag force on a flat plate inclined towards the direction of the fluid flow. Using F for the drag force, ρ for the density, S for the area of the flat plate, V for the flow velocity, and θ for the inclination angle, his law was expressed as

$$F = \rho\, S\, V^2 \sin^2 \theta$$

This equation is incorrect for the calculation of drag in most cases. Drag on a flat plate is closer to being linear with the angle of inclination as opposed to acting quadratically at low angles. The Newton formula can lead one to believe that flight is more difficult than it actually is, due to this overprediction of drag, and thus required thrust, which might have contributed to a delay in human flight. However, it is more correct for a very slender plate when the angle becomes large and flow separation occurs or if the flow speed is supersonic [5].

1.3 Modern Era

In 1738, the Dutch-Swiss mathematician Daniel Bernoulli published *Hydrodynamica*. In this book Bernoulli described the fundamental relationship among pressure, density, and velocity, in particular Bernoulli's principle, which is one method to calculate aerodynamic lift [6]. More general equations of fluid flow – the Euler equations – were published by Leonhard Euler in 1757. The Euler equations were extended to incorporate the effects of viscosity in the first half of the eighteenth century, resulting in the Navier–Stokes equations.

Sir George Cayley is credited as the first person to identify the four aerodynamic forces of flight – weight, lift, drag, and thrust – and the relationships between them [7, 8]. Cayley believed that the drag on a flying machine must be counteracted to enable level flight to occur. He also looked into the nature of aerodynamic shapes with low drag. Among the shapes he investigated were the cross sections of trout. This may appear counterintuitive; however, the bodies of fish are shaped to produce very low resistance as they travel through water. Their cross sections are sometimes very close to that of modern low-drag aerofoils.

Air resistance experiments were carried out by investigators throughout the eighteenth and nineteenth centuries. Drag theories were developed by Jean le Rond d'Alembert [9], Gustav Kirchhoff [10], and Lord Rayleigh [11]. Equations for fluid flow with friction were developed by Claude-Louis Navier [12] and George Gabriel Stokes [13]. To simulate fluid flow, many experiments involved immersing objects in streams of water or simply dropping them off the top of a tall building. Towards the end of this time period, Gustave Eiffel used his Eiffel Tower to assist in the drop testing of flat plates.

A more precise way to measure resistance is to place an object within an artificial, uniform stream of air where the velocity is known. The first person to experiment in this fashion was Francis Herbert Wenham, who in doing so constructed the first wind tunnel in 1871. Wenham was also a member of the first professional organisation dedicated to aeronautics, the Royal Aeronautical Society of the United Kingdom. Objects placed in wind tunnel as models are almost always smaller than in practice, so a method was needed to relate small-scale models to their real-life counterparts. This was achieved with the invention of the dimensionless Reynolds number by Osborne Reynolds [14]. In 1883, Reynolds also experimentally studied laminar to turbulent flow transition.

By the late nineteenth century, two problems were identified before heavier-than-air flight could be realised. The first was the creation of low-drag, high-lift aerodynamic wings. The second problem was how to determine the power needed for sustained flight. During this time, the groundwork was laid down for modern-day fluid dynamics and aerodynamics, with other less scientifically inclined enthusiasts testing various flying machines with little success.

In 1889, Charles Renard, a French aeronautical engineer, became the first person to reasonably predict the power needed for sustained flight [15]. Renard and German physicist Hermann von Helmholtz explored the wing loading (weight-to-wing-area ratio) of birds, eventually concluding that humans could not fly under their own power by attaching wings onto their arms. Otto Lilienthal, following the work of Sir George Cayley, was the first person to become highly successful with glider flights. Lilienthal believed that thin, curved aerofoils would produce high lift and low drag.

Octave Chanute provided a great service to those interested in aerodynamics and flying machines by publishing a book outlining all of the research conducted around the world up to 1997 [16].

1.3.1 Actual Flights

With the information contained in Chanute's book, the personal assistance of Chanute himself, and research carried out in their own wind tunnel, the Wright brothers gained enough knowledge of aerodynamics to fly the first powered aircraft on 17 December 1903. The Wright brothers' flight confirmed or disproved a number of aerodynamic theories. Newton's drag force theory was finally proved incorrect. This first widely publicised flight led to a more organised effort between aviators and scientists, leading the way to modern aerodynamics.

During the time of the first flights, Frederick W. Lanchester [17], Martin Wilhelm Kutta, and Nikolai Zhukovsky independently created theories that connected circulation of a fluid flow to lift. Kutta and Zhukovsky went on to develop a two-dimensional wing theory. Expanding upon the work of Lanchester, Ludwig Prandtl is credited with developing the mathematics [18] behind thin-aerofoil and lifting-line theories and the boundary layers. Prandtl, a professor at the University of Göttingen, instructed many students who would play important roles in the development of aerodynamics, such as Theodore von Karman and Max Munk.

1.3.2 Compressibility Issues

At low speeds, the compressibility of air is not significant in relation to aircraft design, but as the airflow nears and exceeds the speed of sound, a host of new aerodynamic effects become important in the design of aircraft. These effects, often several of them at a time, made it very difficult for World War II-era aircraft to reach speeds much beyond 800 km/h.

Some of the minor effects include changes to the airflow that lead to problems in control. For instance, the P-38 Lightning with its thick high-lift wing had a particular problem in high-speed dives that led to a nose-down condition. Pilots would enter dives and then find that they could no longer control the plane, which continued to nose-down over a distance until it crashed. The problem was remedied by adding a 'dive flap' beneath the wing that altered the centre of pressure distribution so that the wing would not lose its lift [19].

A similar problem affected some models of the Supermarine Spitfire. At high speeds the ailerons could apply more torque than the Spitfire's thin wings could handle, and the entire wing would twist in the opposite direction. This meant that the plane would roll in the direction opposite to that which the pilot intended and led to a number of accidents. Earlier models were not fast enough, for handling this was felt as a problem, and so it was not noticed until later models of Spitfire like the Mk.IX started to appear. This was mitigated by adding considerable torsional rigidity to the wings and was wholly cured when the Mk.XIV was introduced.

The Messerschmitt Bf 109 and Mitsubishi Zero had the exact opposite problem in which the controls became ineffective. At higher speeds the pilot simply could not move the controls because there was too much airflow over the control surfaces. The planes would become difficult to manoeuvre, and at high enough speeds aircraft without this problem could outturn them.

These problems were eventually solved as jet aircraft reached transonic and supersonic speeds. German scientists in World War II experimented with swept wings. Their research was applied on the MiG-15 and F-86 Sabre and bombers such as the B-47 Stratojet used swept wings that delay the onset of shock waves and reduce the drag. The all-flying tailplane which is common on supersonic planes also helps to maintain control near the speed of sound.

Finally, another common problem that fits into this category is flutter. At some speeds, the airflow over the control surfaces will become turbulent, and the controls will start to flutter. If the speed of the fluttering is close to a harmonic of the control's movement, the resonance could break the control off completely. When problems with poor control at high speeds were first encountered, they were addressed by designing a new style of control surface with more power. However, this introduced a new resonant mode, and a number of planes were lost before this was discovered.

All of these effects are often mentioned in conjunction with the term 'compressibility', but in a manner of speaking, they are incorrectly used. From a strictly aerodynamic point of view, the term compressibility should refer only to those side effects arising as a result of the change in the nature of the airflow from incompressible (similar in effect to water) to compressible (acting as a gas) as the speed of sound is approached. There are two effects in particular, wave drag and critical Mach number.

Wave drag is a sudden rise in drag on the aircraft, caused by air building up in front of it. At lower speeds this air has time to 'get out of the way', guided by the air in front of it that is in contact with the aircraft. However, at the speed of sound, this can no longer happen, and the air that was previously following the streamline around the aircraft now hits it directly. The amount of power needed to overcome this effect is considerable. The critical Mach number is the speed at which some of the air passing over the aircraft becomes supersonic.

At the speed of sound, the way that lift is generated changes dramatically, from being dominated by Bernoulli's principle to forces generated by shock waves. Since the air on the top of the wing is travelling faster than on the bottom, due to the Bernoulli effect, at speeds close to the speed of sound, the air on the top of the wing will be accelerated to supersonic level. When this happens the distribution of lift changes dramatically, typically causing a powerful nose-down trim. Since the aircraft normally approach these speeds only in a dive, pilots would report the aircraft attempting to nose-dive into the ground.

An important aspect observed at hypersonic speeds is that the process of dissociation absorbs a great deal of energy in a reversible process. This greatly reduces the thermodynamic temperature of hypersonic gas decelerated near an aerospace vehicle. In transition regions, where this pressure dependent dissociation is incomplete, both the differential, constant-pressure heat capacity, and β (the volume/pressure differential ratio) will greatly increase. The latter has a pronounced effect on vehicle aerodynamics including stability.

1.3.3 Supersonic Speeds

As aircraft began to travel faster, aerodynamicists realised that the density of air began to change as it came into contact with an object, leading to a division of fluid flow into the incompressible and compressible regimes. In compressible aerodynamics, density and pressure both change, which is the basis for calculating the speed of sound. Newton was the first to develop a mathematical model for calculating the speed of sound, but it was not correct until Pierre-Simon Laplace accounted for the molecular behaviour of gases and introduced the heat capacity ratio. The ratio of the flow speed to the speed of sound was named the Mach number after Ernst Mach, who was one of the first to investigate the properties of supersonic flow that included Schlieren photography techniques to visualise the changes in density. William John Macquorn Rankine and Pierre Henri Hugoniot independently developed the theory for flow properties before and after a shock wave. Jakob Ackeret led the initial work on calculating the lift and drag on a supersonic aerofoil [20]. Theodore von Karman and Hugh Latimer Dryden introduced the term transonic to describe flow speeds around Mach 1 where drag increases rapidly. Because of the increase in drag while approaching Mach 1, aerodynamicists and aviators disagreed on whether supersonic flight was achievable.

On 30 September 1935, an exclusive conference was held in Rome with the topic of high velocity flight and the possibility of breaking the sound barrier [21]. Participants included Theodore von Karman, Ludwig Prandtl, Jakob Ackeret, Eastman Jacobs, Adolf Busemann, Geoffrey Ingram Taylor, Gaetano Arturo Crocco, and Enrico Pistolesi. Ackeret presented a design for a supersonic wind tunnel. Busemann gave a presentation on the need for aircraft with swept wings for high speed flight. Eastman Jacobs, working for NACA, presented his optimised aerofoils for high subsonic speeds that led to some of the high-performance American aircraft during World War II. Supersonic propulsion was also discussed. The sound barrier was broken using the Bell X-1 aircraft 12 years later.

By the time the sound barrier was broken, much of the subsonic and low supersonic aerodynamics knowledge had matured. The Cold War (a state of political and military tension after World War II between powers in the Western Bloc (the United States, its NATO allies, and others such as Japan) and powers in the Eastern Bloc (the Soviet Union and its allies in the Warsaw Pact)) fuelled an ever evolving line of high-performance aircraft. Computational fluid dynamics was started as an effort to solve for flow properties around complex objects and has rapidly grown to the point where entire aircraft can be designed using a computer, with wind tunnel tests followed by flight tests to confirm the computer predictions.

With some exceptions, the knowledge of hypersonic aerodynamics has matured between the 1960s and the present decade. Therefore, the goals of an aerodynamicist have shifted from understanding the behaviour of fluid flow to understanding how to engineer a vehicle to interact appropriately with the fluid flow. For example, while the behaviour of hypersonic flow is understood, building a scramjet aircraft to fly at hypersonic speeds has seen very limited success. Along with building a successful scramjet aircraft, the desire to improve the aerodynamic efficiency of current aircraft and propulsion systems will continue to fuel new research in aerodynamics. Nevertheless, there are still important problems in basic aerodynamic theory, such as in predicting transition to turbulence, and the existence and uniqueness of solutions to the Navier–Stokes equations.

1.3.4 Continuity Concept

The foundation of aerodynamic prediction is the continuity assumption. In reality, gases are composed of molecules that collide with one another and solid objects. To derive the equations of aerodynamics, fluid properties such as density and velocity are assumed to be well-defined at infinitely small points and to vary continuously from one point to another. That is, the discrete molecular nature of a gas is ignored. The continuity assumption becomes less valid as a gas becomes more rarefied. In these cases statistical mechanics is a more valid method of solving the problem than continuum aerodynamics. The Knudsen number can be used to guide the choice between statistical mechanics and the continuum formulation of aerodynamics.

1.4 Conservation Laws

Aerodynamic problems are normally solved using conservation of mass, momentum, and energy, referred to as continuity, momentum, and energy equations. The conservation laws can be written in integral or differential form.

1.4.1 Conservation of Mass

If a certain mass of fluid enters a volume, it must either exit the volume or change the mass inside the volume. In fluid dynamics the continuity equation is analogous to Kirchhoff's current law (that is, 'the sum of the currents flowing into a point in a circuit is equal to the sum of the currents flowing out of that same point') in electric circuits. The differential form of the continuity equation is

$$\frac{\partial \rho}{\partial t} + \nabla \cdot (\rho \, u) = 0$$

where ρ is the fluid density, u is a velocity vector, and t is time. Physically the equation also shows that mass is neither created nor destroyed in the control volume. For a steady-state process, the rate at which mass enters the volume is equal to the rate at which it leaves the volume [22]. Consequently, the first term on the left-hand side is then equal to zero. For flow through a tube with one inlet (state 1) and exit (state 2) as shown in Figure 1.1, the continuity equation may be written and solved as

$$\rho_1 u_1 A_1 = \rho_2 u_2 A_2$$

where A is the variable cross-sectional area of the tube at the inlet and exit. For incompressible flows, the density remains constant.

Figure 1.1 Flow through a constant area pipe.

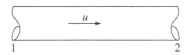

1.4.2 Conservation of Momentum

The momentum equation applies Newton's second law of motion to a control volume in a flow field, whereby force is equal to the time derivative of momentum. Both surface and body forces are accounted for in this equation. For instance, F could be expanded into an expression for the frictional force acting on an internal flow:

$$\frac{Du}{Dt} = F - \frac{\nabla p}{\rho}$$

For the pipe flow in Figure 1.1, control volume analysis gives

$$p_1 A_1 + \rho_1 A_1 u_1^2 + F = p_2 A_2 + \rho_2 A_2 u_2^2$$

where the force F is placed on the left-hand side of the equation, assuming it acts with the flow moving in a left-to-right direction. Depending on the other properties of the flow, the resulting force could be negative that means it acts in the opposite direction as depicted in Figure 1.1. In aerodynamics, air is normally assumed to be a Newtonian fluid, which posits a linear relationship between the shear stress and the rate of strain of the fluid. The equation above is a vector equation: in a three-dimensional flow, it can be expressed as three scalar equations. The conservation of momentum equations are often called the Navier–Stokes equations, while others use the term for the system that includes conversation of mass, conservation of momentum, and conservation of energy.

1.4.3 Conservation of Energy

Although energy can be converted from one form to another, the total energy in a given closed system remains constant:

$$\rho \frac{Dh}{Dt} = \frac{Dp}{Dt} + \nabla \cdot (k \, \nabla T) + \Phi$$

where h is enthalpy, k is the thermal conductivity of the fluid, T is temperature, and Φ is the viscous dissipation function. The viscous dissipation function governs the rate at which mechanical energy of the flow is converted to heat. This term is always positive since, according to the second law of thermodynamics, viscosity cannot add energy to the control volume [23]. The expression on the left-hand side is a material derivative. Again using the pipe flow in Figure1.1, the energy equation in terms of the control volume may be written as

$$\rho_1 u_1 A_1 \left(h_1 + \frac{u_1^2}{2} \right) + \dot{W} + \dot{Q} = \rho_2 u_2 A_2 \left(h_2 + \frac{u_2^2}{2} \right)$$

where the shaft work $\dot{W}$ and heat transfer rate $\dot{Q}$ are assumed to be acting on the flow. They may be positive or negative depending on the problem.

The ideal gas law or another equation of state is often used in conjunction with these equations to form a determined system to solve for the unknown variables.

1.5 Incompressible Aerodynamics

An incompressible flow is characterised by a constant density. While all real fluids are compressible, a flow problem is often considered incompressible if the density changes in the problem have a small effect on the outputs of interest. This is more likely to be true when the flow speeds are significantly lower than the speed of sound. For higher speeds, the flow would encounter significant compressibility as it comes into contact with surfaces and slows down.

1.5.1 Subsonic Flow

Subsonic aerodynamics is the study of fluid motion that is slower than the speed of sound. There are several branches of subsonic flow, but one special case arises when the flow is inviscid, incompressible, and irrotational. This case is called potential flow. For this case, the differential equations used are simplified version of the governing equations of fluid dynamics, thus making a range of quick and easy solutions available to the aerodynamicist [24].

In solving a subsonic problem, one decision to be made by the aerodynamicist is whether to incorporate the effects of compressibility. Compressibility is a description of the amount of change of density in the problem. When the effects of compressibility on the solution are small, the aerodynamicist may choose to assume that density is constant. The problem is then an incompressible low-speed aerodynamics problem. When the density is allowed to vary, the problem is called a compressible problem. In air, compressibility effects are usually ignored when the Mach number in the flow does not exceed 0.3. Above 0.3, the problem should be solved by using compressible aerodynamics [24].

1.6 Compressible Aerodynamics

According to the theory of aerodynamics, a flow is considered to be compressible if its change in density with respect to pressure is more than 5%. This means that – unlike incompressible flow – changes in density must be considered. In general, this is the case where the Mach number in part or all of the flow exceeds 0.3. The Mach 0.3 value is rather arbitrary, but it is used because gas flows with a Mach number below 0.3 demonstrate the changes in density with respect to the change in pressure of less than 5%. Furthermore, a maximum of 5% density change occurs at the stagnation point of an object immersed in the gas flow, and the density changes around the rest of the object will be significantly lower. Transonic, supersonic, and hypersonic flows are all compressible.

1.6.1 Transonic Flow

The term transonic refers to a range of velocities just below and above the local speed of sound (generally taken as Mach 0.8–1.2). It is defined as the range of speeds between the critical Mach number, when some parts of the airflow over an aircraft become supersonic, and a higher speed, typically near Mach 1.2, when all of the airflow is supersonic. Between these speeds some of the airflow is supersonic, and some is not.

1.6.2 Supersonic Flow

Supersonic aerodynamic problems are those involving flow speeds greater than the speed of sound. Calculating the lift on the Concorde during cruise can be an example of a supersonic aerodynamic problem.

Supersonic flow behaves very differently from subsonic flow. Fluids react to differences in pressure; pressure changes in a flow field is 'informed' to the flow by the sound waves. It is known that sound is an infinitesimal pressure difference propagating through a fluid; therefore the speed of sound in that fluid can be considered the fastest speed that 'information' can travel in the flow. This difference most obviously manifests itself in the case of a fluid striking an object. In front of that object, the fluid builds up a stagnation pressure as impact with the object brings the moving fluid to rest. In fluid travelling at subsonic speed, this pressure disturbance can propagate upstream, changing the flow pattern ahead of the object and giving the impression that the fluid 'knows' the object is there and is avoiding it. However, in a supersonic flow, the pressure disturbance cannot propagate upstream. Thus, when the fluid finally does strike the object, it is forced to change its properties – temperature, density, pressure, and Mach number – in an extremely violent and irreversible fashion across a shock wave. The presence of shock waves, along with the compressibility effects of high-velocity fluids, is the central difference between supersonic and subsonic aerodynamic problems.

1.6.3 Hypersonic Flow

In aerodynamics, hypersonic speeds are speeds that are highly supersonic. In the 1970s, the term generally came to refer to speeds of Mach 5 (five times the speed of sound) and above. The hypersonic regime is a subset of the supersonic regime. Hypersonic flow is characterised by high-temperature flow behind a shock wave, viscous interaction, and chemical dissociation of gas.

1.7 Vocabulary

The incompressible and compressible flow regimes produce many associated phenomena, such as boundary layers and turbulence.

1.7.1 Boundary Layers

The concept of a boundary layer is important in many aerodynamic problems. The viscosity and fluid friction in the air are approximated as being significant only in this thin layer. This principle makes aerodynamics much more tractable mathematically.

1.7.2 Turbulence

In aerodynamics, turbulence is characterised by chaotic, stochastic property changes in the flow. This includes low momentum diffusion, high momentum convection, and rapid variation of pressure and velocity in space and time. Flow that is not turbulent is called laminar flow. Incorporating all the characteristics, turbulence may be described as a 'random three-dimensional phenomenon, exhibiting multiplicity of scales, possessing vorticity, and showing very high dissipation'.

1.8 Aerodynamics in Other Fields

Aerodynamics is important in a number of applications other than aerospace engineering. It is a significant factor in any type of vehicle design, including automobiles. It is important in the prediction of forces and moments in sailing. It is used in the design of mechanical components such as hard drive heads. Structural engineers also use aerodynamics, and particularly aeroelasticity, to calculate wind loads in the design of large buildings and bridges. Urban aerodynamics seeks to help town planners and designers improve comfort in outdoor spaces, create urban microclimates, and reduce the effects of urban pollution. The field of environmental aerodynamics studies the ways atmospheric circulation and flight mechanics affect ecosystems. The aerodynamics of internal passages is important in heating/ventilation, gas piping, and automotive engines where detailed flow patterns strongly affect the performance of the engine.

1.9 Essence of Fluid Mechanics

Fluid mechanics has a unique vocabulary associated with it, like any other science. Precise definition of the basic concepts forms the foundation for the proper development of a subject. In this section, all important basic concepts associated with fluid mechanics are discussed. The unit systems that will be used are also reviewed. Careful study of these concepts will be of great value for understanding the topics covered in the following chapters.

1.9.1 Some Basic Facts About Fluid Mechanics

Fluid mechanics may be defined as *the science of fluid flow in which the temperature change associated with flow speed in negligibly small*. Treating any change less than 5% as negligible, in accordance with engineering practice, any fluid flow with speed less than 650 km/h (about 180 m/s), at standard sea-level condition, can be treated as a fluid mechanic stream. This nature of fluid flow renders the energy equation for the flow of perfect gases as passive in the study of fluid mechanics. Thus, the continuity, momentum, and increase of entropy principle are sufficient to solve fluid flows in the regimes of fluid mechanics. Another aspect seen from the definition of fluid mechanics is that a fluid dynamic stream may be incompressible or compressible. That is, when the Mach number is less than 0.3, the changes in density may be neglected, and the stream may be treated as incompressible. In addition, for Mach number more than 0.3, the stream must be treated as compressible. Thus the science of fluid mechanics deals with both incompressible and compressible subsonic flows. Fluid mechanics may also be defined as *the subject dealing with the investigation of the motion and equilibrium of fluids*. It is one of the oldest branches of physics and foundation for the understanding of many essential aspects of applied sciences and engineering. It is a subject of enormous interest in numerous fields such as biology, biomedicine, geophysics, meteorology, physical chemistry, plasma physics, and almost all branches of engineering.

Nearly 200 years ago, man thought of laying down scientific rules to govern the motion of fluids. The rules were used mainly on the flow of water and air to understand them so that he can protect himself from their fury during natural calamities such as cyclone and floods and utilise their power to develop fields like civil engineering and naval architecture. In spite of the common origin, two distinct schools of thought gradually developed. On the one hand, through the concept of an 'ideal fluid', mathematical physicists developed the theoretical science known as *classical hydrodynamics*. On the other hand, realising that idealised theories were of no practical application without empirical correction factors, engineers developed from experimental studies the applied science known as *hydraulics* for the specific fields of irrigation, water supply, river flow control, hydraulic power, and so on. Further, the development of aerospace, chemical, and mechanical engineering during the past few decades and the exploration of space from 1960s have increased the interest in the study of fluid mechanics. Thus, it now ranks as one of the most important basic subjects in engineering science.

The science of fluid flow has been extended into fields like regimes of hypervelocity flight and flow of electrically conducting fluids. This has introduced new fields of interest such as *hypersonic flow* and *magneto-fluid dynamics*. In this connection, it has become essential to combine the knowledge of thermodynamics, heat transfer, mass transfer, electromagnetic theory, and fluid mechanics for the complete understanding of the physical phenomenon involved in any flow process.

Fluid dynamics is one of the rapidly growing basic sciences whose principles find application even in daily life. For instance, the flight of bird in air and the motion of fish in water are governed by the fluid dynamic rules. The design of various types of aircraft and ships is based on the fluid dynamic principles. Even natural phenomena like tornadoes and hurricanes can also be explained by the science of fluid dynamics. In fact, the science of fluid dynamics dealing with such natural phenomena has been developed to such an extent that they can be predicted well in advance. Since the Earth is surrounded by an environment of air and water to a very large extent, almost everything that is happening on earth and its atmosphere is some way or the other associated with the science of fluid dynamics.

The science of fluid motion is referred to as the *Mechanics of Fluids*, an allied subject of the mechanics of solids and engineering materials and built on the same fundamental laws of motion. Therefore, unlike empirical hydraulics, it is based on the physical principles and has close correlation with experimental studies that both compliment and substantiate the fundamental analysis, unlike the classical hydrodynamics that is based purely on mathematical treatment.

For understanding the fluid flows, it is essential to know the properties of fluids. Before discussing the fluid properties, it will be useful if the difference between solids and fluids is understood. From the basic studies on physics, it is known that solids, liquids, and gases are the three states of matter. In general, liquids and gases are called fluids. It can be shown that this division into solid and fluid states constitutes a natural grouping of matter from the standpoint of internal stresses and strains in elastic media, that is, the stress in a linear elastic solid is proportional to strain, while the stress in a fluid is proportional to its time rate of strain. In fact, among fluids themselves, only a group of fluids exhibit the above said stress–strain relation, and they are called *Newtonian fluids*. The abovementioned behaviour of solids and fluids may also be expressed in a simple way, as follows:

- When a force is applied to a solid, deformation will be produced in the solid. If the force per unit area, viz. stress, is less than the yield stress, that is, within the proportional limit of the material, the deformation disappears when the applied force is removed. If the stress is more than the yield stress of the material, it will acquire a permanent setting or even break.
- If the shearing force is applied to a fluid, it will deform continuously as long as the force is acting on it, regardless of the magnitude of the force.

The above difference in behaviour between solids and liquids can be explained by their molecular properties. The existence of very strong intermolecular attractive forces in solids lends them rigidity. The forces are comparatively weaker and extremely small in liquids and gases, respectively. This characteristic enables the liquid molecules to move freely within a liquid mass while still maintaining a close proximity to one another, whereas the gas molecules have freedom to the extent of completely filling any space allotted to them.

The study of fluid flows can be classified into the following divisions:

1. *Fluid statics*: Dealing with fluid elements at rest with respect to one another and thus free of shearing stress. The static pressure distribution in a fluid and on bodies immersed in a fluid can be determined from a static analysis.

2. *Kinematics of fluids*: Dealing with translation, rotation, and rate of deformation of fluid particles. This analysis is useful in determining methods to describe the motion of fluid particles and analysing flow patterns. However, the velocity and acceleration of fluid particles cannot be obtained from kinematic study alone, since the interaction of fluid particles with one another makes the fluid a disturbed medium.

3. *Dynamic analysis*: Dealing with the determination of the effects of the fluid and its surroundings on the motion of the fluid. This involves the consideration of forces acting on the fluid particles in motion with respect to one another. Since there is a relative motion between fluid particles, shearing forces must be taken into consideration in the dynamic analysis.

1.9.2 Fluids and the Continuum

Fluid flows may be modelled on either microscopic or macroscopic level. The macroscopic model regards the fluid as a *continuum*, and the description is in terms of the variations of macroscopic velocity, density, pressure, and temperature with distance and time. On the other hand, the microscopic or molecular model recognises the particulate structure of a fluid as a myriad of discrete molecules and ideally provides information on the position and velocity of every molecule at all times.

The description of a fluid motion essentially involves a study of the behaviour of all the discrete molecules that constitute the fluid. In liquids, the strong intermolecular cohesive forces make the fluid behave as a continuous mass of substance, and, therefore, these forces need to be analysed by the molecular theory. Under normal conditions of pressure and temperatures, even gases have large number of molecules in unit volume (for example, under normal conditions, for most gases, the molecular density is 2.7×10^{25} molecules/m^3), and, therefore, they also can be treated as a continuous mass of substance by considering the average effects of all the molecules within the gas. Such a fluid model is called continuum.

The continuum approach must be used only where it may yield reasonably correct results. For instance, this approach breaks down when the mean free path, *the average distance travelled by the molecules between successive collisions*, of the molecules is of the same order of magnitude as the smallest significant length in the problem being investigated. Under such circumstances, detection of meaningful, gross manifestation of molecules is not possible. The action of each molecule or group of molecules is then of significance and must be treated accordingly.

To understand this, it is essential to investigate the action of a gas on an elemental area inside a closed container. Even if the quantity of gas is assumed to be small, innumerable collisions of molecules on the surface result in the gross, non-time-dependent manifestations of force. That is, the gas acts like a continuous substance. On the other hand, if only a tiny amount of gas is kept in the container so that the mean free path is of the same order of magnitude as the sides of the area element, an erratic activity is experienced, as individual or groups of molecules bombard the surface. This cannot be treated as a constant force; one must deal with an erratic force variation, as shown graphically in Figure 1.2.

A continuous distribution of mass cannot exhibit this kind of variation. Thus, it is seen that in the first case, the continuum approach would be applicable, but in the second case the continuum approach would be questionable. In Figure 1.2, it is clear that, when the mean free path is large in comparison with some characteristic length, the gas cannot be considered continuous and must be analysed on the molecular scale. The mean free path, the statistical average distance that molecules travel between collisions, of atmospheric air is between 50 and 70 nm. The other factor that influences the molecular activities of gas is the elapsed time between collisions.

Figure 1.2 Force variation with time.

This time must be sufficiently small so that the random statistical nature of the molecular activity is preserved.

This book deals only with continuous fluids. Further, it will be assumed that the elastic properties are the same at all points in the fluid and are identical in all directions from any specified point. These stipulations make the fluid both *homogeneous* and *isotropic*.

1.9.3 Dimension and Units

In fluid dynamics, mostly the gross, measurable molecular manifestations such as pressure and density and other equally important measurable abstract entities such as length and time will be dealt with. These manifestations that are characteristics of the behaviour of a particular fluid, and not of the manner of flow, may be called *fluid properties*. Density and viscosity are example of fluid properties. To adequately discuss these properties, a consistent set of standard units must be defined. Table 1.1 gives the common system of units and their symbol.

In this text, throughout we shall use the SI system of units. However, other systems of units are equally applicable to all the equations.[2]

1.9.4 Law of Dimensional Homogeneity

This law states, '*an analytically derived equation representing a physical phenomenon must be valid for all system of units*'. Thus, the equation for the frequency of a simple pendulum, $f = (1/2\pi) \sqrt{g/l}$, is properly stated for any system of units. This explains why all natural phenomena proceed completely in accordance with man-made units, and hence fundamental equations representing such events should have validity for any system of units. Thus, the fundamental equations of physics are dimensionally homogeneous, and consequently all relations derived from these equations must

Table 1.1 Common systems of units.

Quantity	Unit	SI	CGS	FPS	MKS
Mass	Kilogram	kg	g	lb	kg
Length	Metre	m	cm	ft	m
Time	Second	s	s	s	s
Force	Newton	$N = (kg\,m)/s$	dyn	pdl	kgf
Temperature	Kelvin	K	°C	°F	°C

---

2 Conversion table for SI units to American units is given in Appendix A.

also be dimensionally homogeneous. For this to occur under all systems of units, it is necessary that each grouping in an equation has the same dimensional representation.

Examine the following dimensional representation of an equation:

$$L = T^2 + T$$

where L denotes length and T the time. Changing the units of length from feet to metres will change the value of the left-hand side while not affecting the right-hand side, thus making the equation invalid in the new system of units. Dimensionally homogeneous equations only will be considered in this book.

1.9.5 The Perfect Gas Equation of State

Gases are basically divided into two broad categories:

Perfect gas is that in which intermolecular forces are negligible.
Real gas is that in which intermolecular forces are important and must be accounted for.

In this book, we are concerned only with the fluids that can be regarded as perfect gases. For perfect gases, the kinetic theory of gases indicates that there exists a simple relation between pressure, specific volume, and absolute temperature. For a perfect gas at equilibrium, this relation has the following form:

$$\boxed{pv = RT} \tag{1.1}$$

This is known as the *perfect gas state equation*. This equation is also called ideal gas equation of state or simply the *ideal gas relation*, and a gas that obeys this relation is called an ideal gas. However, it is essential to understand the difference between perfect and ideal gases. A perfect gas is one that is calorically perfect. That is, its specific heats at constant pressure (c_p) and constant volume (c_v) are constants and independent of temperature. A perfect gas can be viscous or inviscid. Also, perfect gas flow may be incompressible or compressible. However, an *ideal gas* is assumed to be *inviscid* and *incompressible*. Therefore, we can state that ideal gas is a special case of perfect gas. In Eq. (1.1), p is the absolute pressure, T is the absolute temperature, v is the specific volume, and R is the gas constant. The gas constant R is different for each gas and is determined from

$$R = \frac{R_u}{M} \quad \left[\text{kJ/(kg K) or (kPa m}^3\text{)/(kg K)} \right]$$

where R_u is the universal gas constant and M is the molar mass (also called molecular weight). The universal gas constant R_u is same for all substances, and its value is

$$R_u = \begin{cases} 8.314 \text{ kJ/(kmol K)} \\ 8.314 \text{ (kPa m}^3\text{)/(kmol K)} \\ 0.083\,14 \text{ (bar m}^3\text{)/(kmol K)} \\ 1.986 \text{ Btu/(lb mol R)} \\ 10.73 \text{ (psia ft}^3\text{)/(lb mol R)} \\ 1545.00 \text{ (ft lbf)/(lb mol R)} \end{cases}$$

The *molar mass M* can be simplify defined as the mass of 1 mol of a substance in grams or the mass of 1 kmol in kilograms. It is essential to realise that an ideal gas is *an imaginary substance* that obeys the relation $pv = RT$. It has been experimentally observed that the ideal gas relation given above closely approximates the p–v–T behaviour of real gases at low densities. At low pressures and high temperatures too, a gas behaves as an ideal gas, because the intermolecular force becomes less significant compared with the kinetic energy of the molecules. However, at lower temperatures and higher pressures, the ideal gas model tends to fail since the intermolecular force and molecular size become important at these conditions. The ideal gas model fails for most heavy gases, such as refrigerants, and for gases with strong intermolecular forces, such as water vapour. At high pressures, the volume of a real gas is often considerably greater than that of an ideal gas. At low temperatures, the pressure of a real gas is usually considerably less than that of an ideal gas. It should be noted that the ideal gas model does not describe phase transition.

Basically state equation relates the pressure, density, and temperature of a substance. Although many substances are complex in behaviour, the experience shows that most gases of practical interest at moderate pressure and temperature are well represented by the perfect gas state equation

$$p = \rho \, R \, T$$

Although no real substance behaves exactly as an ideal or perfect gas, the perfect gas equation is in error by less than 1% for air at room temperature for pressures up to 30 atm. For air at 1 atm, the equation is in error by less than 1% for temperatures as low as 140 K.

In the range of practical interest, many familiar gases such as air, nitrogen, oxygen, hydrogen, helium, argon, neon, krypton, and even heavier gases such as carbon dioxide can be treated as ideal gases with negligible error (often less than 1%). However, dense gases such as water vapour in steam power plants and refrigerant vapour in refrigerator should not be treated as ideal gases.

1.9.6 Regimes of Fluid Mechanics

Based on the flow properties that characterise the physical situation, the flows are classified into various types as follows.

1.9.6.1 Ideal Fluid Flow

It is only an imaginary situation where the fluid is assumed to be inviscid or nonviscous and incompressible. Therefore, there is no tangential force between adjacent fluid layers. An extensive mathematical theory is available for ideal fluid flow. Although the theory of ideal fluid flow fails to account for viscous and compressibility effects in actual fluid flow processes, it gives reasonably reliable results in the calculation of lift, induced drag, and wave motion for gas flow at low velocity and for water flow. This branch of fluid dynamics is called *classical hydrodynamics*.

1.9.6.2 Viscous Incompressible Flow

The theory of viscous incompressible flow assumes fluid density to be constant. It finds widespread application in the flow of liquids and the flow of air at low velocity. The phenomena involving viscous forces, flow separation, and eddy flows are studied with the help of this theory.

1.9.6.3 Gas Dynamics

Gas dynamics may be defined as *the science of flow field where a change in pressure is accompanied by density and temperature changes*. The theory of gas dynamics deals with the dynamics and thermodynamics of the flow of compressible fluids. Based on the dimensionless velocity, namely, Mach number (M), defined as *the ratio of flow velocity and the local speed of sound*, gas dynamics can be further divided into the fields of study commonly referred to as *subsonic* ($M < 1$), *transonic* ($M \approx 1$), *supersonic* ($1 < M < 5$), and *hypersonic* ($5 < M < 40$) gas dynamics. The classification between supersonic and hypersonic flows is arbitrary because it depends on the shape of the body over which the fluid flows. In a gas dynamic flow, a change in velocity is accompanied by a change of density and temperature. Thus, gas dynamics is essentially a combination of fluid mechanics and thermodynamics. It is referred to as *aerothermodynamics* when it deals with the aerodynamic forces and moments and the heat distribution of a vehicle that flies at hypersonic speeds. For gas dynamics, in addition to Mach number, Prandtl number and the ratio of specific heats $\gamma\ (= c_p/c_v)$ also play an important role being control parameters. The Prandtl number, *Pr*, which is *the ratio of kinematic viscosity and thermal diffusivity, is a measure of the relative importance of velocity and heat conduction.*

Regardless of speed ranges, the theory of gas dynamics can be divided into two parts, inviscid gas dynamics and viscous gas dynamics. The inviscid theory is important in the calculation of nozzle characteristics, shock waves, lift, and wave drag of a body, while the viscous theory is applicable to the calculation of skin friction and heat transfer characteristics of a body moving through a gas, such as atmospheric air.

1.9.6.4 Rarefied Gas Dynamics

The concept of continuum fails when the mean free path of fluid molecules are comparable with some characteristic geometrical parameter in the flow field. A dimensionless parameter, Knudsen number, *Kn*, defined as *the ratio of mean free path to a characteristic length*, aptly describes the degree of departure from continuum flow. Based on the Knudsen number, the flow regimes are grouped as follows:

- *Continuum (Kn* < 0.01): All equations of viscous compressible flow are applicable in this regime. The no-slip boundary condition is valid.
- *Slip flow* (0.01 < *Kn* < 0.1): Here again the continuum fluid dynamic analysis is applicable provided the slip boundary conditions are employed. That is, the no-slip boundary condition of continuum flow, dictating zero velocity at the surface of an object kept in the flow, is not valid. The fluid molecules move (slip) with a finite velocity, called slip velocity, at the boundary.
- *Transition flow* (0.01 < *Kn* < 5): In this regime of flow, the fluid cannot be treated as continuum. At the same time, it cannot be treated as a free molecular flow since such a flow demands the intermolecular force of attraction to be negligible. Hence, it is a flow regime between continuum and free molecular. The kinetic theory of gases must be employed to adequately describe this flow.
- *Free molecular flow (Kn* > 5): In this regime of flow, the fluid molecules are so widely dispersed that the intermolecular forces can be neglected.
  All these regimes of rarefied gas dynamics or super aerodynamics are encountered at high altitudes, where the molecular density is very low. This branch of fluid flow is also called *low-density flow.*
- *Magneto-fluid mechanics*: The subject of magneto-fluid mechanics is an extension of fluid mechanics, with thermodynamics, mechanics, materials, and electrical sciences. This branch was initiated by astrophysicists. Other names that are used to refer to this discipline are magneto-hydrodynamics, magneto-gas dynamics, and hydro-magnetics.

Magneto-fluid mechanics is the study of the motion of an electrically charged conducting fluid in the presence of a magnetic field. The motion of a electricity conducting fluid in the magnetic field will induce electric currents in the fluid, thereby modifying the field. The flow field will also be modified by the mechanical forces produced by it. The interaction between the field and the motion makes magneto-fluid dynamics analysis difficult.

A gas at normal and moderately high temperatures is a non-conductor. However, at very high temperatures, of the order of 10 000 K and above, thermal excitation sets in. This leads to dissociation and ionisation. Ionised gas is called plasma, which is an electrically conducting medium. Electrically conducting fluids are encountered in engineering problems like re-entry of missiles and spacecraft, plasma jet, controlled fusion research, and magneto-hydrodynamic generator.

1.9.6.5 Flow of Multicomponent Mixtures

This field is simply an extension of basic fluid mechanics. The analysis of flow of homogeneous fluid consisting of single species, termed basic fluid mechanics, is extended to study the flow of chemically reacting component mixtures, made of more than one species. All the three transports, namely, the momentum transport, energy transport, and mass transport, are considered in this study, unlike the basic fluid mechanics where only transport of momentum and energy are considered.

1.9.6.6 Non-Newtonian Fluid Flow

Fluids for which the stress is not proportional to time rate of strain are called *non-Newtonian fluids*. Such fluids show a nonlinear dependence of shearing stress on velocity gradient. Examples of non-Newtonian fluids are honey, printer's ink, paste, and tar.

In 1872, an international meeting in France proposed a treaty called the Metric Convention, which was signed in 1875 by 17 countries including the United States. It was an important step over British systems because its use of base 10 is the foundation of our number system. Problems still remained because even the metric countries differed in their use of kilo-pounds instead of dynes or newton, kilograms instead of grams, or calories instead of joule. To standardise the metric system, a General Conferences of Weights and Measures attended by 40 countries in 1960 proposed the International System of Units (SI units). The SI units become more and more popular throughout the world, and it is expected to replace all other systems of units in due course.

The International System of Units (abbreviated SI from French: Le Système international d'unités) is the modern form of the metric system and is the world's most widely used system of units, used in both everyday commerce and science. It comprises a coherent system of units of measurement built around seven base units, 22 named and an indeterminate number of unnamed coherent derived units, and a set of prefixes that act as decimal-based multipliers. It is part of the International System of Quantities.

As we saw, there are four primary dimensions in fluid mechanics from which all other dimensions can be derived, namely, mass, length, time, and temperature. These dimensions and their units are given in Table 1.1. The braces around a symbol like L mean 'the dimensions of' length. All other variables in fluid mechanics can be expressed in terms of $\{M\}$, $\{L\}$, $\{T\}$, and $\{Q\}$. For instance, velocity has the dimensions $\{LT^{-1}\}$. One of the interesting secondary dimensions is that of force, F, which is directly related to mass, length, and time. By Newton's second law,

$$F = m\,a$$

where m is mass and a is acceleration. From this relation it is seen that, dimensionally, $F = MLT^{-2}$. A constant of proportionality that will figure in the above force relation is avoided by defining the force unit exactly in terms of the primary units. Thus, the Newton can be defined as

$$1 \text{ newton of force} = 1 \text{ N} \equiv 1 \text{ (kg m)/ s}^2$$

Some of the secondary variables that are often used in the study of the fluid dynamics along with their dimensions in terms of the primary dimensions are given in Table 1.2.

Table 1.2   Units of some secondary variables.

Dimension	SI unit
Acceleration $\{LT^{-2}\}$	m/s^2
Angular velocity $\{T^{-1}\}$	s^{-1}
Area $\{L^2\}$	m^2
Density $\{ML^{-3}\}$	kg/m^3
Energy, heat, and work $\{ML^2T^{-2}\}$	$J = N\,m$
Power $\{ML^2T^{-3}\}$	$W = J/s$
Pressure or stress $\{ML^{-1}T^{-2}\}$	$Pa = N/m^2$
Velocity $\{LT^{-1}\}$	m/s
Viscosity $\{ML^{-1}T^{-1}\}$	$kg/(m\,s)$
Volume $\{L^3\}$	m^3

1.10 Summary

Aerodynamics is the study of forces and the resulting motion of objects through the air.

Aerodynamics may also be described as a branch of dynamics concerned with studying the motion of air, particularly when it interacts with a solid object.

Aerodynamics allows the definition and solution of equations for the conservation of mass, momentum, and energy in air. The use of aerodynamics through mathematical analysis, empirical approximations, wind tunnel experimentation, and computer simulations forms the scientific basis for heavier-than-air flight and a number of other technologies.

External aerodynamics is the study of flow around solid objects of various shapes. Evaluating the lift and drag on an airplane or the shock waves that form in front of the nose of a rocket are examples of external aerodynamics. Internal aerodynamics is the study of flow through passages in solid objects. For instance, internal aerodynamics encompasses the study of the airflow through a jet engine.

A problem is called *subsonic* if all the speeds in the problem are less than the speed of sound, *transonic* if speeds both below and above the speed of sound are present, *supersonic* if the flow speed is greater than the speed of sound, and *hypersonic* if the flow speed is more than five times the speed of sound.

The influence of viscosity in the flow dictates a third classification.

Sir Isaac Newton was the first to develop a theory of air resistance, making him one of the first aerodynamicists. Newton considered that drag was due to the dimensions of a body, the density of the fluid, and the velocity raised to the second power. The drag force F was expressed as

$$F = \rho\, S\, V^2 \sin^2 \theta$$

This equation is incorrect for the calculation of drag in most cases.

The Dutch-Swiss mathematician Daniel Bernoulli published *Hydrodynamica*. In this book, Bernoulli described the fundamental relationship among pressure, density, and velocity, in particular Bernoulli's principle, which is one method to calculate aerodynamic lift. More general equations of fluid flow – the Euler equations – were published by Leonhard Euler. The Euler equations were extended to incorporate the effects of viscosity, resulting in the Navier–Stokes equations.

Sir George Cayley is credited as the first person to identify the four aerodynamic forces of flight – weight, lift, drag, and thrust. Cayley also looked in to the nature of aerodynamic shapes with low drag.

Drag theories were developed by Jean le Rond d'Alembert, Gustav Kirchhoff, and Lord Rayleigh. Equations for fluid flow with friction were developed by Claude-Louis Navier and George Gabriel Stokes.

A precise way to measure resistance is to place an object within an artificial, uniform stream of air where the velocity is known. The first person to experiment in this fashion was Francis Herbert Wenham, who in doing so constructed the first wind tunnel.

Objects placed in wind tunnel as models are almost always smaller than in practice, so a method was needed to relate small-scale models to their real-life counterparts. This was achieved with the invention of the dimensionless Reynolds number by Osborne Reynolds. Reynolds also experimentally studied laminar to turbulent flow transition.

Charles Renard was the first person to reasonably predict the power needed for sustained flight. Renard and German physicist Hermann von Helmholtz explored the wing loading.

Sir George Cayley was the first person to become highly successful with glider flights.

The Wright brothers' flight confirmed or disproved a number of aerodynamic theories. Newton's drag force theory was finally proved incorrect. This first widely publicised flight led to a more organised effort between aviators and scientists, leading the way to modern aerodynamics.

During the time of the first flights, Frederick W. Lanchester, Martin Wilhelm Kutta, and Nikolai Zhukovsky independently created theories that connected circulation of a fluid flow to lift. Kutta and Zhukovsky went on to develop a two-dimensional wing theory. Expanding upon the work of Lanchester, Ludwig Prandtl is credited with developing the mathematics behind thin-aerofoil and lifting-line theories and the boundary layers.

At low speeds the compressibility of air is not significant in relation to aircraft design, but as the airflow nears and exceeds the speed of sound, a host of new aerodynamic effects become important in the design of aircraft.

Wave drag is a sudden rise in drag on the aircraft, caused by air building up in front of it.

The critical Mach number is the speed at which some of the air passing over the aircraft's wing becomes supersonic.

At the speed of sound, the way that lift is generated changes dramatically, from being dominated by Bernoulli's principle to forces generated by shock waves.

It is found that the process of dissociation absorbs a great deal of energy in a reversible process. This greatly reduces the thermodynamic temperature of hypersonic gas decelerated near an aerospace vehicle.

The ratio of the flow speed to the speed of sound was named the Mach number.

William John Macquorn Rankine and Pierre Henri Hugoniot independently developed the theory for flow properties before and after a shock wave. Jakob Ackeret led the initial work on calculating the lift and drag on a supersonic aerofoil.

Theodore von Karman and Hugh Latimer Dryden introduced the term transonic to describe flow speeds around Mach 1 where drag increases rapidly.

Ackeret presented a design for a supersonic wind tunnel. Busemann gave a presentation on the need for aircraft with swept wings for high-speed flight. Eastman Jacobs, working for NACA, presented his optimised aerofoils for high subsonic speeds that led to some of the high-performance American aircraft during World War II.

The knowledge of hypersonic aerodynamics has matured between the 1960s and the present decade.

The foundation of aerodynamic prediction is the continuity assumption. The continuity assumption becomes less valid as a gas becomes more rarefied.

Aerodynamic problems are normally solved using conservation of mass, momentum, and energy, referred to as continuity, momentum, and energy equations.

The differential form of the continuity equation is

$$\frac{\partial \rho}{\partial t} + \nabla \cdot (\rho \, u) = 0$$

Momentum equation applies Newton's second law of motion to a control volume in a flow field, whereby force is equal to the time derivative of momentum. Both surface and body forces are accounted for in this equation.

Although energy can be converted from one form to another, the total energy in a given closed system remains constant:

$$\rho \frac{Dh}{Dt} = \frac{Dp}{Dt} + \nabla \cdot (k \, \nabla T) + \Phi$$

where h is enthalpy, k is the thermal conductivity of the fluid, T is temperature, and Φ is the viscous dissipation function.

The energy equation in terms of the control volume may be written as

$$\rho_1 u_1 A_1 \left(h_1 + \frac{u_1^2}{2} \right) + \dot{W} + \dot{Q} = \rho_2 u_2 A_2 \left(h_2 + \frac{u_2^2}{2} \right)$$

where the shaft work $\dot{W}$ and heat transfer rate $\dot{Q}$ are assumed to be acting on the flow. They may be positive (to the flow from the surroundings) or negative (to the surroundings from the flow) depending on the problem.

An incompressible flow is characterised by a constant density. While all real fluids are compressible, a flow problem is often considered incompressible if the density changes in the problem have a small effect on the outputs of interest.

Subsonic aerodynamics is the study of fluid motion that is everywhere much slower than the speed of sound through the fluid or gas. There are several branches of subsonic flow, but one special case arises when the flow is inviscid, incompressible, and irrotational.

According to the theory of aerodynamics, a flow is considered to be compressible if its change in density with respect to pressure is more than 5%. This is the case where the Mach number in part or all of the flow exceeds 0.3.

The term transonic refers to a range of velocities just below and above the local speed of sound (generally taken as Mach 0.8–1.2).

Supersonic aerodynamic problems are those involving flow speeds greater than the speed of sound.

In aerodynamics, hypersonic speeds are speeds that are highly supersonic. The term generally came to refer to speeds of Mach 5 (five times the speed of sound) and above. Hypersonic flow is characterised by high-temperature flow behind a shock wave, viscous interaction, and chemical dissociation of gas.

The concept of a boundary layer is important in many aerodynamic problems. The viscosity and fluid friction in the air is approximated as being significant only in this thin layer. This principle makes aerodynamics much more tractable mathematically.

Turbulence may be described as a 'random three-dimensional phenomenon, exhibiting multiplicity of scales, possessing vorticity, and showing very high dissipation'.

Aerodynamics is important in a number of applications other than aerospace engineering. It is a significant factor in any type of vehicle design, including automobiles. It is important in the prediction of forces and moments in sailing. It is used in the design of mechanical components such as hard drive heads. Structural engineers also use aerodynamics, and particularly aeroelasticity, to calculate wind loads in the design of large buildings and bridges. Urban aerodynamics seeks to help town planners and designers improve comfort in outdoor spaces, create urban microclimates, and reduce the effects of urban pollution. The field of environmental aerodynamics studies the ways atmospheric circulation and flight mechanics affect ecosystems. The aerodynamics of internal passages is important in heating/ventilation, gas piping, and automotive engines where detailed flow patterns strongly affect the performance of the engine.

2

International Standard Atmosphere

The International Standard Atmosphere (ISA) is an atmospheric model of how the pressure, temperature, density, and viscosity of the Earth's atmosphere change over a wide range of altitudes. It has been established to provide a common reference for temperature and pressure and consists of tables of values at various altitudes, plus some formulae by which those values were derived. The International Organisation for Standardisation (ISO) publishes the ISA as an international standard, ISO 2533:1975 [25]. Other standards organisations, such as the International Civil Aviation Organisation (ICAO) and the US government, publish extensions or subsets of the same atmospheric model under their own standards-making authority.

The ISA model divides the atmosphere into layers with linear temperature distributions [26]. The other values are computed from basic physical constants and relationships. Thus, the standard consists of a table of values at various altitudes, plus some formulas by which those values were derived. For example, at sea level, the standard gives a pressure of 1013.25 hPa (1 atm), a temperature of 15 °C, and an initial lapse rate of 6.5 °C/km (roughly 2 °C1000 ft). The tabulation continues to 11 km where the pressure has fallen to 226.32 hPa and the temperature to −56.5 °C. Between 11 and 20 km, the temperature remains constant [27, 28].

2.1 Layers in the ISA

Different layers in the atmosphere and the corresponding values of height, temperature lapse rates, temperatures, and pressures are listed in Table 2.1.

2.1.1 ICAO Standard Atmosphere

The International Civil Aviation Organisation (ICAO) published their ICAO Standard Atmosphere as Doc 7488-CD in 1993. It has the same model as the ISA but extends the altitude coverage to 80 km (262 500 ft) [29]. The ICAO Standard Atmosphere does not contain water vapour.

Some of the values defined by ICAO are listed in Table 2.2.

As this is a standard, we will not always encounter these conditions outside of a laboratory, but many aviation standards and flying rules are based on this, altimetry being a major one. The standard is very useful in meteorology for comparing against actual values.

2.1.2 Temperature Modelling

The temperature variation in the standard atmosphere is given in Figure 2.1. Temperature decreases with altitude at a constant rate of −6.5 °C/1000 m (−1.98 °C/1000 ft) up to the tropopause.

Introduction to Aerospace Engineering: Basic Principles of Flight, First Edition. Ethirajan Rathakrishnan.
© 2021 John Wiley & Sons, Inc. Published 2021 by John Wiley & Sons, Inc.
Companion Website: www.wiley.com/go/Rathakrishnan/IntroductiontoAerospaceEngineering

Table 2.1 Different layers in standard atmosphere, along with the temperature and pressure there.

Layer	Level name	Base geopotential height h (km)	Base geometric height z (km)	Lapse rate (°C/km)	Base temperature T (°C)	Base pressure p (Pa)
0	Troposphere	0.0	0.0	6.5	+15.0	101 325
1	Tropopause	11.000	11.019	+0.0	56.5	22 632
2	Stratosphere	20.000	20.063	+1.0	56.5	5 474.9
3	Stratosphere	32.000	32.162	+2.8	44.5	868.02
4	Stratopause	47.000	47.350	+0.0	2.5	110.91
5	Mesosphere	51.000	51.413	2.8	2.5	66.939
6	Mesosphere	71.000	71.802	2.0	58.5	3.9564
7	Mesopause	84.852	86.000		86.2	0.3734

Table 2.2 Some representative values of pressure and temperature in the ICAO Standard Atmosphere.

Height, km (ft)	Temperature (°C)	Pressure (hPa)	Lapse rate (°C/1000 ft)
0 mean sea level (0)	15.0	1 013.25	1.98 (tropospheric)
11 (36 000)	56.5	226.00	0.00 (stratospheric)
20 (65 000)	56.5	54.70	+0.3 (stratospheric)
32 (105 000)	44.5	8.68	

The standard tropopause altitude is 11 000 m (36 089 ft). Therefore, the air that is considered as a perfect gas in the ISA model presents the following characteristics within the troposphere:

$$T = T_0 - 6.5 \left(\frac{h}{1000} \right) \tag{2.1}$$

where the altitude h is in metres. The temperature remains at a constant value of $-56.5\,°C$ (216.65 K) from the tropopause up to 20 000 m (65 600 ft).

This ISA model is used as a reference to compare real atmospheric conditions and the corresponding engine/aircraft performance. The atmospheric conditions will therefore be expressed as ISA $+/-$ at a given flight level.

Example 2.1

Determine the atmospheric temperature at 9500 m altitude.

Solution

Given: $h = 9500$ m.

The local temperature, by Eq. (2.1), is

$$T = T_0 - 6.5 \left(\frac{h}{1000} \right)$$

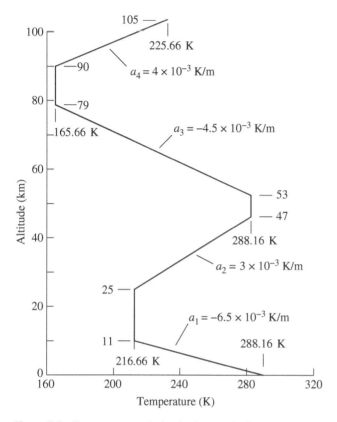

Figure 2.1 Temperature variation in the standard atmosphere.

where $T_0 = 15\,°C$ is the temperature at sea level, that is, at $h = 0$. Therefore, the standard temperature at 9500 m is

$$T = T_0 - 6.5 \times \left(\frac{9500}{1000}\right)$$
$$= 288.15 - 61.75$$
$$= 226.4 \text{ K}$$
$$= 226.4 - 273.15$$
$$= \boxed{-46.75\,°C}$$

2.2 Pressure Modelling

To calculate the standard pressure p at a given altitude, the temperature is assumed standard, and the air is assumed as a perfect gas. The altitude obtained from the measurement of the pressure is called pressure altitude (PA). Table 2.3 gives the variation of the pressure altitude as a function of the pressure. The last column of Table 2.3 shows corresponding flight levels for the given pressure altitudes.

Table 2.3 Pressure altitude versus pressure.[a]

Pressure (hPa)	Pressure altitude (m)
200	11 784
250	10 363
300	9 164
500	5 574
850	4 813
1 013	0

a) hPa is hectopascal; 1hPa = 100 Pa.

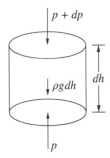

Figure 2.2 An element in atmosphere.

The pressure variations for the ISA can be calculated by using the hydrostatic equation, perfect gas law, and temperature lapse rate equation. The hydrostatic equation for a column of air (Figure 2.2) is

$$dp = -\rho g dh \tag{2.2}$$

The state equation for a perfect gas is

$$p = \rho R T \tag{2.3}$$

where R is the gas constant for air. Dividing Eq. (2.2) by (2.3), we have

$$\frac{dp}{p} = -\frac{\rho g dh}{\rho R T} = -\left(\frac{g}{RT}\right) dh \tag{2.4}$$

The relationship between the pressure at a troposphere altitude and sea-level pressure can be obtained by integrating Eq. (2.4) between $h_0 = 0$ and h:

$$\int_{p_0}^{p} \frac{dp}{p} = -\frac{g}{R} \int_{h_0=0}^{h} \frac{dh}{T_0 - 0.0065\,h}$$

Integrating, we get

$$p = p_0 \left(1 - 0.0065\,\frac{h}{T_0}\right)^{5.2561} \tag{2.5}$$

In this equation, T_0 is in Kelvin and h is in metres.

2.2.1 Pressure Above the Tropopause

For the altitudes above the tropopause, the temperature is constant, so integrating Eq. (2.4) from the tropopause to an altitude above the tropopause is

$$\int_{p_{11}}^{p} \frac{dp}{p} = -\frac{g}{RT_{11}} \int_{h_{11}=11\ 000}^{h} dh$$

This gives

$$p = p_{11} e^{-g(h-h_{11})/(RT_{11})} \tag{2.6}$$

where the parameters with subscript '11' correspond to the values at the tropopause and $p_{11} = 226.32\,\text{hPa}$, $T_{11} = 216.65\ \text{K}$, and $h_{11} = 11\ 000$ m.

2.3 Density Modelling

Using the pressure and temperature values calculated for a given altitude, the standard density can easily be calculated from the perfect gas state equation

$$\rho = \frac{p}{RT} \tag{2.7}$$

The ISA parameters such as temperature, pressure, density, and speed of sound are listed as a function of the altitude in Table 2.4.

Table 2.4 Properties of standard atmosphere.

h (m)	T (K)	p (kN/m^2)	ρ (kg/m^3)	a (m/s)
0	288.150	1.01320e+02	1.22500e+00	3.40263e+02
100	287.500	1.00125e+02	1.21329e+00	3.39879e+02
200	286.850	9.89416e+01	1.20167e+00	3.39494e+02
300	286.200	9.77695e+01	1.19013e+00	3.39109e+02
400	285.550	9.66086e+01	1.17867e+00	3.38724e+02
500	284.900	9.54589e+01	1.16730e+00	3.38338e+02
600	284.250	9.43204e+01	1.15602e+00	3.37952e+02
700	283.600	9.31928e+01	1.14482e+00	3.37566e+02
800	282.950	9.20762e+01	1.13370e+00	3.37178e+02
900	282.300	9.09704e+01	1.12266e+00	3.36791e+02
1000	281.650	8.98755e+01	1.11171e+00	3.36403e+02
1100	281.000	8.87912e+01	1.10084e+00	3.36015e+02
1200	280.350	8.77175e+01	1.09005e+00	3.35626e+02
1300	279.700	8.66544e+01	1.07934e+00	3.35236e+02
1400	279.050	8.56017e+01	1.06871e+00	3.34847e+02

(Continued)

Table 2.4 (Continued)

h (m)	*T* (K)	*p* (kN/m^2)	ρ (kg/m^3)	*a* (m/s)
1500	278.400	8.45594e+01	1.05816e+00	3.34456e+02
1600	277.750	8.35274e+01	1.04769e+00	3.34066e+02
1700	277.100	8.25056e+01	1.03730e+00	3.33675e+02
1800	276.450	8.14940e+01	1.02700e+00	3.33283e+02
1900	275.800	8.04924e+01	1.01676e+00	3.32891e+02
2000	275.150	7.95008e+01	1.00661e+00	3.32499e+02
2100	274.500	7.85191e+01	9.96535e−01	3.32106e+02
2200	273.850	7.75473e+01	9.86537e−01	3.31712e+02
2300	273.200	7.65852e+01	9.76616e−01	3.31318e+02
2400	272.550	7.56329e+01	9.66772e−01	3.30924e+02
2500	271.900	7.46901e+01	9.57003e−01	3.30529e+02
2600	271.250	7.37569e+01	9.47310e−01	3.30134e+02
2700	270.600	7.28331e+01	9.37693e−01	3.29738e+02
2800	269.950	7.19187e+01	9.28150e−01	3.29342e+02
2900	269.300	7.10136e+01	9.18681e−01	3.28945e+02
3000	268.650	7.01178e+01	9.09287e−01	3.28548e+02
3100	268.000	6.92312e+01	8.99967e−01	3.28150e+02
3200	267.350	6.83536e+01	8.90719e−01	3.27752e+02
3300	266.700	6.74851e+01	8.81545e−01	3.27353e+02
3400	266.050	6.66255e+01	8.72443e−01	3.26954e+02
3500	265.400	6.57749e+01	8.63413e−01	3.26554e+02
3600	264.750	6.49330e+01	8.54455e−01	3.26154e+02
3700	264.100	6.40999e+01	8.45568e−01	3.25754e+02
3800	263.450	6.32754e+01	8.36752e−01	3.25352e+02
3900	262.800	6.24596e+01	8.28006e−01	3.24951e+02
4000	262.150	6.16523e+01	8.19330e−01	3.24549e+02
4100	261.500	6.08535e+01	8.10724e−01	3.24146e+02
4200	260.850	6.00631e+01	8.02188e−01	3.23743e+02
4300	260.200	5.92810e+01	7.93720e−01	3.23339e+02
4400	259.550	5.85071e+01	7.85321e−01	3.22935e+02
4500	258.900	5.77415e+01	7.76990e−01	3.22531e+02
4600	258.250	5.69840e+01	7.68727e−01	3.22126e+02
4700	257.600	5.62346e+01	7.60531e−01	3.21720e+02
4800	256.950	5.54931e+01	7.52402e−01	3.21314e+02
4900	256.300	5.47596e+01	7.44340e−01	3.20907e+02

Table 2.4 (Continued)

h (m)	T (K)	p (kN/m²)	ρ (kg/m³)	a (m/s)
5000	255.650	5.40340e+01	7.36344e−01	3.20500e+02
5100	255.000	5.33162e+01	7.28414e−01	3.20092e+02
5200	254.350	5.26061e+01	7.20550e−01	3.19684e+02
5300	253.700	5.19037e+01	7.12750e−01	3.19275e+02
5400	253.050	5.12089e+01	7.05016e−01	3.18866e+02
5500	252.400	5.05217e+01	6.97345e−01	3.18456e+02
5600	251.750	4.98419e+01	6.89739e−01	3.18046e+02
5700	251.100	4.91696e+01	6.82196e−01	3.17635e+02
5800	250.450	4.85046e+01	6.74717e−01	3.17224e+02
5900	249.800	4.78469e+01	6.67300e−01	3.16812e+02
6000	249.150	4.71965e+01	6.59946e−01	3.16399e+02
6100	248.500	4.65532e+01	6.52654e−01	3.15986e+02
6200	247.850	4.59171e+01	6.45424e−01	3.15573e+02
6300	247.200	4.52880e+01	6.38255e−01	3.15159e+02
6400	246.550	4.46659e+01	6.31148e−01	3.14744e+02
6500	245.900	4.40508e+01	6.24101e−01	3.14329e+02
6600	245.250	4.34425e+01	6.17114e−01	3.13913e+02
6700	244.600	4.28411e+01	6.10187e−01	3.13497e+02
6800	243.950	4.22464e+01	6.03320e−01	3.13080e+02
6900	243.300	4.16584e+01	5.96513e−01	3.12663e+02
7000	242.650	4.10770e+01	5.89764e−01	3.12245e+02
7100	242.000	4.05023e+01	5.83074e−01	3.11826e+02
7200	241.350	3.99340e+01	5.76442e−01	3.11407e+02
7300	240.700	3.93723e+01	5.69867e−01	3.10988e+02
7400	240.050	3.88169e+01	5.63351e−01	3.10567e+02
7500	239.400	3.82679e+01	5.56891e−01	3.10147e+02
7600	238.750	3.77252e+01	5.50488e−01	3.09725e+02
7700	238.100	3.71888e+01	5.44142e−01	3.09303e+02
7800	237.450	3.66586e+01	5.37852e−01	3.08881e+02
7900	236.800	3.61345e+01	5.31618e−01	3.08458e+02

(Continued)

Table 2.4 (Continued)

h (m)	T (K)	p (kN/m^2)	ρ (kg/m^3)	a (m/s)
8000	236.150	3.56164e+01	5.25439e−01	3.08034e+02
8100	235.500	3.51044e+01	5.19315e−01	3.07610e+02
8200	234.850	3.45984e+01	5.13246e−01	3.07185e+02
8300	234.200	3.40983e+01	5.07231e−01	3.06760e+02
8400	233.550	3.36041e+01	5.01270e−01	3.06334e+02
8500	232.900	3.31157e+01	4.95363e−01	3.05907e+02
8600	232.250	3.26330e+01	4.89510e−01	3.05480e+02
8700	231.600	3.21561e+01	4.83709e−01	3.05052e+02
8800	230.950	3.16848e+01	4.77961e−01	3.04624e+02
8900	230.300	3.12191e+01	4.72266e−01	3.04195e+02
9000	229.650	3.07590e+01	4.66623e−01	3.03765e+02
9100	229.000	3.03044e+01	4.61031e−01	3.03335e+02
9200	228.350	2.98553e+01	4.55491e−01	3.02904e+02
9300	227.700	2.94115e+01	4.50002e−01	3.02473e+02
9400	227.050	2.89731e+01	4.44563e−01	3.02041e+02
9500	226.400	2.85400e+01	4.39175e−01	3.01608e+02
9600	225.750	2.81122e+01	4.33837e−01	3.01175e+02
9700	225.100	2.76896e+01	4.28549e−01	3.00741e+02
9800	224.450	2.72721e+01	4.23311e−01	3.00307e+02
9900	223.800	2.68598e+01	4.18121e−01	2.99871e+02
10000	223.150	2.64525e+01	4.12980e−01	2.99436e+02
10100	222.500	2.60502e+01	4.07888e−01	2.98999e+02
10200	221.850	2.56529e+01	4.02844e−01	2.98562e+02
10300	221.200	2.52605e+01	3.97848e−01	2.98124e+02
10400	220.550	2.48730e+01	3.92899e−01	2.97686e+02
10500	219.900	2.44903e+01	3.87997e−01	2.97247e+02
10600	219.250	2.41124e+01	3.83143e−01	2.96807e+02
10700	218.600	2.37392e+01	3.78335e−01	2.96367e+02
10800	217.950	2.33708e+01	3.73573e−01	2.95926e+02
10900	217.300	2.30069e+01	3.68857e−01	2.95485e+02

Table 2.4 (Continued)

h (m)	T (K)	p (kN/m²)	ρ (kg/m³)	a (m/s)
11100	216.650	2.22935e+01	3.58492e−01	2.95042e+02
11200	216.650	2.19449e+01	3.52886e−01	2.95042e+02
11300	216.650	2.16018e+01	3.47368e−01	2.95042e+02
11400	216.650	2.12640e+01	3.41936e−01	2.95042e+02
11500	216.650	2.09314e+01	3.36589e−01	2.95042e+02
11600	216.650	2.06041e+01	3.31326e−01	2.95042e+02
11700	216.650	2.02819e+01	3.26145e−01	2.95042e+02
11800	216.650	1.99648e+01	3.21044e−01	2.95042e+02
11900	216.650	1.96526e+01	3.16024e−01	2.95042e+02
12000	216.650	1.93453e+01	3.11082e−01	2.95042e+02
12100	216.650	1.90427e+01	3.06218e−01	2.95042e+02
12200	216.650	1.87450e+01	3.01429e−01	2.95042e+02
12300	216.650	1.84518e+01	2.96716e−01	2.95042e+02
12400	216.650	1.81633e+01	2.92076e−01	2.95042e+02
12500	216.650	1.78793e+01	2.87508e−01	2.95042e+02
12600	216.650	1.75997e+01	2.83013e−01	2.95042e+02
12700	216.650	1.73245e+01	2.78587e−01	2.95042e+02
12800	216.650	1.70536e+01	2.74230e−01	2.95042e+02
12900	216.650	1.67869e+01	2.69942e−01	2.95042e+02
13000	216.650	1.65244e+01	2.65721e−01	2.95042e+02
13100	216.650	1.62660e+01	2.61566e−01	2.95042e+02
13200	216.650	1.60116e+01	2.57476e−01	2.95042e+02
13300	216.650	1.57612e+01	2.53449e−01	2.95042e+02
13400	216.650	1.55148e+01	2.49486e−01	2.95042e+02
13500	216.650	1.52722e+01	2.45585e−01	2.95042e+02
13600	216.650	1.50333e+01	2.41744e−01	2.95042e+02
13700	216.650	1.47983e+01	2.37964e−01	2.95042e+02
13800	216.650	1.45668e+01	2.34243e−01	2.95042e+02
13900	216.650	1.43391e+01	2.30580e−01	2.95042e+02

(Continued)

Table 2.4 (Continued)

h (m)	T (K)	p (kN/m²)	ρ (kg/m³)	a (m/s)
14000	216.650	1.41148e+01	2.26974e−01	2.95042e+02
14100	216.650	1.38941e+01	2.23425e−01	2.95042e+02
14200	216.650	1.36768e+01	2.19931e−01	2.95042e+02
14300	216.650	1.34630e+01	2.16492e−01	2.95042e+02
14400	216.650	1.32524e+01	2.13106e−01	2.95042e+02
14500	216.650	1.30452e+01	2.09774e−01	2.95042e+02
14600	216.650	1.28412e+01	2.06494e−01	2.95042e+02
14700	216.650	1.26404e+01	2.03265e−01	2.95042e+02
14800	216.650	1.24427e+01	2.00086e−01	2.95042e+02
14900	216.650	1.22482e+01	1.96957e−01	2.95042e+02
15000	216.650	1.20566e+01	1.93877e−01	2.95042e+02
15100	216.650	1.18681e+01	1.90846e−01	2.95042e+02
15200	216.650	1.16825e+01	1.87861e−01	2.95042e+02
15300	216.650	1.14998e+01	1.84924e−01	2.95042e+02
15400	216.650	1.13200e+01	1.82032e−01	2.95042e+02
15500	216.650	1.11430e+01	1.79185e−01	2.95042e+02
15600	216.650	1.09687e+01	1.76383e−01	2.95042e+02
15700	216.650	1.07972e+01	1.73625e−01	2.95042e+02
15800	216.650	1.06284e+01	1.70910e−01	2.95042e+02
15900	216.650	1.04622e+01	1.68237e−01	2.95042e+02
16000	216.650	1.02986e+01	1.65607e−01	2.95042e+02
16100	216.650	1.01375e+01	1.63017e−01	2.95042e+02
16200	216.650	9.97900e+00	1.60468e−01	2.95042e+02
16300	216.650	9.82295e+00	1.57958e−01	2.95042e+02
16400	216.650	9.66934e+00	1.55488e−01	2.95042e+02
16500	216.650	9.51814e+00	1.53057e−01	2.95042e+02
16600	216.650	9.36930e+00	1.50663e−01	2.95042e+02
16700	216.650	9.22279e+00	1.48307e−01	2.95042e+02
16800	216.650	9.07857e+00	1.45988e−01	2.95042e+02
16900	216.650	8.93660e+00	1.43705e−01	2.95042e+02

Table 2.4 (Continued)

h (m)	T (K)	p (kN/m^2)	ρ (kg/m^3)	a (m/s)
17 000	216.650	8.79685e+00	1.41458e−01	2.95042e+02
17 100	216.650	8.65929e+00	1.39246e−01	2.95042e+02
17 200	216.650	8.52388e+00	1.37069e−01	2.95042e+02
17 300	216.650	8.39059e+00	1.34925e−01	2.95042e+02
17 400	216.650	8.25938e+00	1.32815e−01	2.95042e+02
17 500	216.650	8.13023e+00	1.30738e−01	2.95042e+02
17 600	216.650	8.00309e+00	1.28694e−01	2.95042e+02
17 700	216.650	7.87794e+00	1.26682e−01	2.95042e+02
17 800	216.650	7.75475e+00	1.24701e−01	2.95042e+02
17 900	216.650	7.63348e+00	1.22751e−01	2.95042e+02
18 000	216.650	7.51412e+00	1.20831e−01	2.95042e+02
18 100	216.650	7.39661e+00	1.18942e−01	2.95042e+02
18 200	216.650	7.28095e+00	1.17082e−01	2.95042e+02
18 300	216.650	7.16709e+00	1.15251e−01	2.95042e+02
18 400	216.650	7.05502e+00	1.13449e−01	2.95042e+02
18 500	216.650	6.94470e+00	1.11675e−01	2.95042e+02
18 600	216.650	6.83610e+00	1.09928e−01	2.95042e+02
18 700	216.650	6.72920e+00	1.08209e−01	2.95042e+02
18 800	216.650	6.62397e+00	1.06517e−01	2.95042e+02
18 900	216.650	6.52039e+00	1.04851e−01	2.95042e+02
19 000	216.650	6.41843e+00	1.03212e−01	2.95042e+02
19 100	216.650	6.31806e+00	1.01598e−01	2.95042e+02
19 200	216.650	6.21926e+00	1.00009e−01	2.95042e+02
19 300	216.650	6.12200e+00	9.84452e−02	2.95042e+02
19 400	216.650	6.02627e+00	9.69058e−02	2.95042e+02
19 500	216.650	5.93204e+00	9.53904e−02	2.95042e+02
19 600	216.650	5.83927e+00	9.38987e−02	2.95042e+02
19 700	216.650	5.74796e+00	9.24304e−02	2.95042e+02
19 800	216.650	5.65808e+00	9.09850e−02	2.95042e+02
19 900	216.650	5.56960e+00	8.95622e−02	2.95042e+02
20 000	216.650	5.48251e+00	8.81617e−02	2.95042e+02

2.3.1 Other Standard Atmospheres

The US Standard Atmosphere is a set of models that define values for atmospheric temperature, density, pressure, and other properties over a wide range of altitudes. The first model, based on an existing international standard, was published in 1958 by the US Committee on Extension to the Standard Atmosphere [30] and was updated in 1962 [31], 1966 [32], and 1976 [33]. The US Standard Atmosphere, ISA, and World Meteorological Organisation (WMO) standard atmospheres are the same as the ISO ISA for altitudes up to 32 km [34, 35].

NRLMSISE-00 is an empirical, global model of the Earth's atmosphere from ground to space. It models the temperatures and densities of the atmosphere's components. A primary use of this model is to aid predictions of satellite orbital decay due to atmospheric drag.

The standard conditions for temperature and pressure are a model of gas temperature and pressure used in chemistry.

2.4 Relative Density

The relative density σ, defined as the ratio of density at the relevant altitude to the density at the sea level in the ISA, is an important quantity in aerodynamics. The value of density at sea level in the ISA, or standard density, as it is often called, is

$$
\begin{aligned}
\rho_0 &= \frac{p_0}{RT_0} \\
&= \frac{101\ 325}{287 \times 288.15} \\
&= 1.225 \ \text{kg/m}^3
\end{aligned}
$$

The relative density is

$$
\boxed{\sigma = \frac{\rho}{\rho_0}} \tag{2.8}
$$

Note that here subscript '0' is used to refer sea-level condition and not the stagnation state.

2.5 Altimeter

Altimeter is an instrument to indicate the altitude at which an aircraft flies. All aircraft are required to be fitted with one or more altimeters to indicate the altitude at which the aircraft is flying. All altimeters consist of an evacuated aneroid capsule fixed by one side to the instrument casing and the other side of the capsule being free to move as the capsule expands or contracts. The movement of the free side of the capsule is communicated by a gearing system to two or three needles that move over a circular scale calibrated in feet or metres. Atmospheric static pressure is applied to the interior of the instrument case through a static tube mounted on the exterior of the aircraft. The equilibrium distention of the capsule is determined by the pressure difference between the instrument case and the evacuated capsule and thus is directly related to the atmospheric pressure. Therefore, the altitude indicated on the scale by the needle is related to the pressure of the atmosphere in which the aircraft is flying.

2.6 Summary

The ISA is an atmospheric model of how the pressure, temperature, density, and viscosity of the Earth's atmosphere change over a wide range of altitudes.

The ISA model divides the atmosphere into layers with linear temperature distributions.

At sea level the standard gives a pressure of 1013.25 hPa (1 atm), a temperature of 15 °C, and an initial lapse rate of 6.5 °C/km (roughly 2 °C/1000 ft).

As this is a standard, we will not always encounter these conditions outside of a laboratory, but many aviation standards and flying rules are based on this, altimetry being a major one. The standard is very useful in meteorology for comparing against actual values.

Temperature decreases with altitude at a constant rate of −6.5 °C/1000 m (−1.98 °C/1000 ft) up to the tropopause.

The temperature remains at a constant value of −56.5 °C (216.65 K) from the tropopause up to 20 000 m (65 600 ft).

To calculate the standard pressure p at a given altitude, the temperature is assumed standard, and the air is assumed as a perfect gas. The altitude obtained from the measurement of the pressure is called pressure altitude.

The pressure variations for the ISA can be calculated by using the hydrostatic equation, perfect gas law, and temperature lapse rate equation.

The state equation for a perfect gas is

$$p = \rho RT$$

where R is the gas constant for air.

Using the pressure and temperature values calculated for a given altitude, the standard density can easily be calculated from the perfect gas state equation:

$$\rho = \frac{p}{RT}$$

The US Standard Atmosphere is a set of models that define values for atmospheric temperature, density, pressure, and other properties over a wide range of altitudes.

NRLMSISE-00 is an empirical, global model of the Earth's atmosphere from ground to space.

The relative density σ, defined as the ratio of density at the relevant altitude to the density at the sea level in the ISA, is an important quantity in aerodynamics.

Altimeter is an instrument to indicate the altitude at which an aircraft flies. All aircraft are required to be fitted with one or more altimeters to indicate the altitude at which the aircraft is flying.

Problems

2.1 Determine the altitude at which the local temperature is −30 °C.

2.2 The Mach number of an aircraft cruising at an altitude is 0.7. If the true air speed is 225 m/s, find the altitude.

3

Aircraft Configurations

Aircraft configurations, which describe the aerodynamic layout or specific components of an aircraft, vary widely. Aircraft configurations, in general, include fuselage, tail, and power plant configurations. This type of aircraft is referred to as fixed wing airplanes. Another category of flying machine with rotating wing is called rotary-winged aircraft or simply rotorcraft. Helicopter, cyclogyro/cyclocopter, autogyro (or gyrocopter, gyroplane, or rotaplane), gyrodyne, and rotor kite (or gyro glider) are all rotary-winged aircraft. An aircraft is a machine that is able to fly by gaining support from the air or, in general, the atmosphere of a planet. It counters the force of gravity by using either static lift or dynamic lift of an aerofoil or in a few cases the downward thrust from jet engines. The human activity that surrounds aircraft is called aviation. Crewed aircraft are flown by an onboard pilot, but unmanned aerial vehicles may be remotely controlled or self-controlled by onboard computers. Aircraft may be classified by different criteria, such as lift type, propulsion, usage, and others.

In general, powered aircraft are classified as propeller aircraft, jet aircraft, and rotorcraft. The key parts of an aircraft are generally divided into the following three categories:

1. *The structure*: The main load-bearing elements and associated equipment.
2. *The propulsion system (if it is powered)*: The power source and associated equipment.
3. *The avionics*: The control, navigation, and communication systems, usually electrical in nature.

3.1 Structure

Structural design varies widely between different types of aircraft. For flying machines such as paraglider, the structure is made of flexible materials that act in tension and rely on aerodynamic pressure to hold their shape. A balloon similarly relies on internal gas pressure but may have a rigid basket or gondola slung below it to carry its payload. Early aircraft, including airships, often employed flexible doped aircraft fabric covering to give a reasonably smooth aeroshell stretched over a rigid frame. Later aircraft employed semi-monocoque techniques, where the skin of the aircraft is stiff enough to share much of the flight loads. In a true monocoque design, there is no internal structure left.

Lighter-than-air types are characterised by one or more gasbags, typically with a supporting structure of flexible cables or a rigid framework called its hull. Other elements such as engines or a gondola may also be attached to the supporting structure.

Heavier-than-air types are characterised by one or more wings and a central fuselage. The fuselage typically also carries a tail or empennage for stability and control, and an undercarriage for

Introduction to Aerospace Engineering: Basic Principles of Flight, First Edition. Ethirajan Rathakrishnan.
© 2021 John Wiley & Sons, Inc. Published 2021 by John Wiley & Sons, Inc.
Companion Website: www.wiley.com/go/Rathakrishnan/IntroductiontoAerospaceEngineering

take-off and landing. Engines may be located on the fuselage or wings. On a fixed-wing aircraft, the wings are rigidly attached to the fuselage, while on a rotorcraft the wings are attached to a rotating vertical shaft. Smaller designs sometimes use flexible materials for part or all of the structure, held in place either by a rigid frame or by air pressure. The fixed parts of the structure comprise the airframe.

3.2 Propulsion

Gliders are heavier-than-air aircraft that do not employ propulsion once airborne. Take-off may be by launching forward and downward from a high location or by pulling into the air on a towline, either by a ground-based winch or vehicle or by a powered 'tug' aircraft. For a glider to maintain its forward airspeed and lift, it must descend in relation to the air (but not necessarily in relation to the ground). Many gliders can 'soar' – gain height from updrafts such as thermal currents. The first practical, controllable example was designed and built by the British scientist and pioneer George Cayley, whom many recognise as the first aeronautical engineer. Common examples of gliders are sailplanes, hang gliders, and paragliders.

Propeller aircraft use one or more propellers (airscrews) to create thrust in the forward direction. The propeller is usually mounted in front of the power source in tractor configuration but can be mounted behind in pusher configuration. Variations of propeller layout include contra-rotating propellers and ducted fans.

Many kinds of power plant have been used to drive propellers. Early airships used manpower or steam engines. The more practical internal combustion piston engine was used for virtually all fixed-wing aircraft until World War II and is still used in many smaller aircraft. Some types use turbine engines to drive a propeller in the form of a turboprop or propfan. Unmanned aircraft and models have also used power sources such as electric motors and rubber bands.

Jet aircraft use air-breathing jet engines, which take in air, burn fuel with it in a combustion chamber, and accelerate the exhaust rearwards to provide thrust. Schematic diagram of a typical aircraft with air-breathing engines is shown in Figure 3.1.

Turbojet and turbofan engines use a spinning turbine to drive one or more fans, which provide additional thrust. An afterburner may be used to inject extra fuel into the hot exhaust, especially on military 'fast jets'. The use of a turbine is not absolutely necessary: other designs include the pulse jet and ramjet. These mechanically simple designs cannot work when stationary, so the aircraft must be launched to flying speed by some other method. Other variants have also been used, including the motorjet and hybrids such as the Pratt & Whitney J58, which can convert between turbojet and ramjet operation.

Compared with propellers, jet engines can provide much higher thrust, higher speeds, and higher efficiency, above about 40 000 ft (12 000 m). They are also much more fuel efficient than rockets. As a consequence, nearly all large, high-speed, or high-altitude aircraft use jet engines.

Some rotorcraft, such as helicopters, have a powered rotary wing or rotor, where the rotor disc can be angled slightly forward so that a proportion of its lift is directed forward. The rotor may, like a propeller, be powered by a variety of methods such as a piston engine or turbine. Experiments have also used jet nozzles at the rotor blade tips.

Rocket-powered aircraft have occasionally been experimented with, and the Messerschmitt Komet fighter even saw action in World War II. Since then, they have been restricted to research aircraft, such as the North American X-15, which travelled up into space where air-breathing engines cannot work (rockets carry their own oxidant). Rockets have more often been used

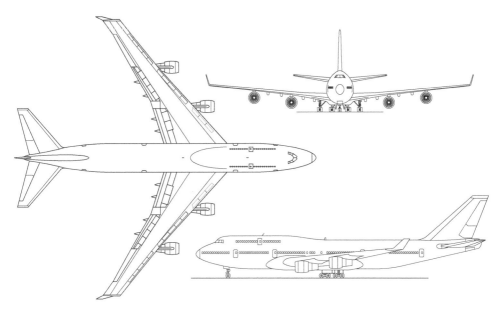

Figure 3.1 Schematic views of Boeing 747 aircraft.

as a supplement to the main power plant, typically for the rocket-assisted take-off of heavily loaded aircraft, but also to provide high-speed dash capability in some hybrid designs such as the Saunders-Roe SR.53.

The ornithopter obtains thrust by flapping its wings. It has found practical use in a model hawk to freeze prey animals into stillness so that they can be captured and in toy birds.

3.3 Summary

Aircraft configurations, which describes the aerodynamic layout or specific components of an aircraft, vary widely. Aircraft configurations, in general, include fuselage, tail, and power plant configuration. This type of aircraft is referred to as fixed wing airplanes. Another category of flying machine with rotating wing is called rotary-winged aircraft or simply rotorcraft. Helicopter, cyclogyro/cyclocopter, autogyro (or gyrocopter, gyroplane, or rotaplane), gyrodyne, and rotor kite (or gyro glider) are all rotary-winged aircraft. An aircraft is a machine that is able to fly by gaining support from the air, or, in general, the atmosphere of a planet.

Powered aircraft are classified as propeller aircraft, jet aircraft, and rotorcraft. The key parts of an aircraft are generally divided into the following three categories:

1. *The structure*: The main load-bearing elements and associated equipment.
2. *The propulsion system* (if it is powered): The power source and associated equipment.
3. *The avionics*: The control, navigation, and communication systems, usually electrical in nature.

Lighter-than-air types are characterised by one or more gasbags, typically with a supporting structure of flexible cables or a rigid framework called its hull.

Heavier-than-air types are characterised by one or more wings and a central fuselage.

Gliders are heavier-than-air aircraft that do not employ propulsion once airborne.

Propeller aircraft use one or more propellers (airscrews) to create thrust in the forward direction.

Jet aircraft use air-breathing jet engines, which take in air, burn fuel with it in a combustion chamber, and accelerate the exhaust rearwards to provide thrust.

Turbojet and turbofan engines use a spinning turbine to drive one or more fans, which provide additional thrust. An afterburner may be used to inject extra fuel into the hot exhaust, especially on military 'fast jets'. The use of a turbine is not absolutely necessary: other designs include the pulse jet and ramjet.

Compared with propellers, jet engines can provide much higher thrust, higher speeds, and higher efficiency, above about 40 000 ft (12 000 m). They are also much more fuel efficient than rockets.

Some rotorcraft, such as helicopters, have a powered rotary wing or rotor, where the rotor disc can be angled slightly forward so that a proportion of its lift is directed forwards.

The ornithopter obtains thrust by flapping its wings. It has found practical use in a model hawk to freeze prey animals into stillness so that they can be captured and in toy birds.

4

Low-Speed Aerofoils

4.1 Introduction

Aerofoil characteristics such as the lift, drag, and pitching moments are discussed by considering the basic features of the pressure distribution around it, as well as the aerodynamic forces acting on it, treating the aerofoil as a two-dimensional body.

4.2 The Aerofoil

An aerofoil is a *streamlined body that would experience the maximum aerodynamic efficiency (that is, maximum lift-to-drag ratio) compared with any other body under identical flow conditions*. To satisfy this characteristics, the body is shaped to have a geometry, as shown in Figure 4.1.

The geometry is so shaped to ensure streamlined flow as far as possible, from the leading edge. Indeed, it is preferable to have streamlined flow from the leading edge to the trailing edge. The leading edge is rounded to ensure smooth flow. The trailing edge is sharp so that the Kutta condition (a body with a sharp trailing edge that is moving through a fluid will create about itself a *circulation* of sufficient strength to hold the rear stagnation point at the trailing edge) may be satisfied, the wake is kept thin, and any region of separated flow is restricted to as small as possible. These features help to achieve high lift-to-drag ratio.

The major geometrical parameters of an aerofoil are the following:

- *Chord line*: The *chord line* is defined as the shortest (straight) line connecting the leading and trailing edges. In other words, this is a reference straight line drawn through the centres of curvatures of the leading and trailing edges.
- *Chord*: Chord c is the shortest distance between the leading and trailing edges. That is, chord is the distance between the points of intersection of the chord line with the leading and trailing edges.
- *Maximum thickness*: Maximum thickness t_{max} is measured normal to the chord line. It is usually expressed as a fraction of the chord, t/c, termed the thickness-to-chord ratio. The t/c is generally about 12–14% for subsonic aerofoils and as low as 3% or 4% for supersonic aerofoils.
- *Camber line*: Camber line is the bisector of the profile thickness from the chord of the profile, as illustrated in Figure 4.1.
- *Camber*: Camber is the maximum deviation d of the *camber line* from the chord.

Introduction to Aerospace Engineering: Basic Principles of Flight, First Edition. Ethirajan Rathakrishnan.
© 2021 John Wiley & Sons, Inc. Published 2021 by John Wiley & Sons, Inc.
Companion Website: www.wiley.com/go/Rathakrishnan/IntroductiontoAerospaceEngineering

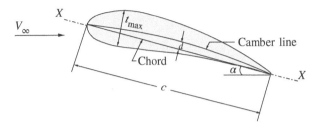

Figure 4.1   Section of an aerofoil.

- The distance of the point of maximum thickness and maximum camber aft of the leading edge are usually expressed as a fraction of the chord.
- The attitude of the aerofoil is expressed by the angle between the chord line and the freestream velocity vector. This angle, denoted by α, illustrated in Figure 4.1, is called the *angle of incidence* or *angle of attack*.

4.3   Aerodynamic Forces and Moments on an Aerofoil

Aerodynamic force acting on an aircraft is the force due to the pressure distribution around it, caused by the motion of the aircraft. Thus, the gravity does not enter into the specification of aerodynamic force. Assuming the motion of the aircraft to be steady without rotation, the aerodynamic force on the wing or on the complete aircraft may be expected to depend on the forward speed V, air density ρ, speed of sound a, and kinematic viscosity v of the environment in which it is flying and the total length l of the aircraft.

The aerodynamic force acts along a line whose intersection with the chord line is called the *centre of pressure cp* of the aerofoil, as shown in Figure 4.2. The aerodynamic force may be resolved into two components, one normal to and one parallel to the freestream flow direction. These components are, respectively, called *lift* and *drag* denoted by L and D.

An alternative method of representing these forces is to shift the aerodynamic force to some other point on the chord line away from cp. This kind of representation results in a moment M, in addition to lift L and drag D, as illustrated in Figure 4.3.

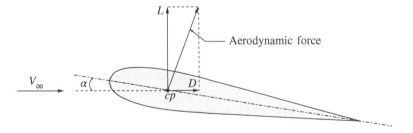

Figure 4.2   Aerodynamic force on an aerofoil.

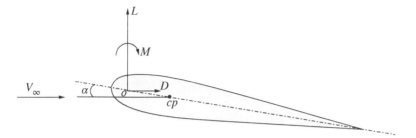

Figure 4.3 Lift, drag, and pitching moment acting on an aerofoil.

4.4 Force and Moment Coefficients

The important aerodynamic forces and moment associated with a flying machine, such as an aircraft, are the lift L, drag D, and pitching moment M. The lift and drag forces can be expressed as dimensionless numbers, popularly known as *lift coefficient C_L* and *drag coefficient C_D* by dividing L and D with $\frac{1}{2}\rho V^2 S$. Thus,

$$\boxed{C_L = \frac{L}{\frac{1}{2}\rho V^2 S}} \tag{4.1a}$$

$$\boxed{C_D = \frac{D}{\frac{1}{2}\rho V^2 S}} \tag{4.1b}$$

where S is the wing planform area, normal to the lift. This area usually includes the projected area of fuselage area in the direction normal to the lift, in addition to the wing area.

The pitching moment, which is the moment of the aerodynamic force about an axis perpendicular to the plane of symmetry, will depend on the particular axis chosen. Denoting the pitching moment about the chosen axis by M (note that M is also used for denoting Mach number, which is the ratio of local flow speed and local speed of sound), we define the *pitching moment coefficient* as

$$\boxed{C_M = \frac{M}{\frac{1}{2}\rho V^2 S c}} \tag{4.1c}$$

where c is the chord of the wing. The pitching moment can also be expressed as

$$C_M = \frac{L \times d}{\frac{1}{2}\rho V^2 S c}$$

where d is the distance between points o and cp in Figure 4.3. Therefore,

$$C_M = -\frac{d}{c} \frac{L}{\frac{1}{2}\rho V^2 S}$$

$$= -\frac{d}{c} C_L$$

This shows that C_M need not be zero when $C_L = 0$. At zero lift also there is a pitching moment, which is nose down, that is, positive for a wing with positive camber. The implication is that as $C_L \to 0$ $cp \to \infty$, that is, the centre of pressure cp moves infinitely far downstream.

4.5 Pressure Distribution

For an aerofoil in an airstream, there are local changes in velocity round the aerofoil, and consequently changes the static pressure, in accordance with the Bernoulli's theorem. This distribution of pressure results in the lift, pitching moment, and form drag[1] acting on the aerofoil and the position of its centre of pressure. The pressure coefficient, which is the difference between the local and freestream pressure in dimensionless form, is usually expressed as

$$C_p = \frac{p - p_\infty}{\frac{1}{2}\rho_\infty V_\infty^2} = 1 - \left(\frac{V}{V_\infty}\right)^2$$

where p and V are the local pressure and velocity, respectively, and p_∞, ρ_∞, and V_∞, respectively, are the freestream pressure, density, and velocity.

From the expression for C_p, the following can be inferred:

- At the stagnation point, C_p has the limiting maximum value of 1. However, there is no overall limit for its lower value.
- A positive pressure coefficient implies a pressure greater than the freestream value; a negative pressure coefficient implies a pressure less than the freestream value. The negative pressure is referred to as a *suction*. At this stage, it is essential to note that what is referred to as suction pressure is relative to some reference, such as the freestream pressure in the present case, and there is no question of suction in any absolute sense, that is, there can be no negative pressure.

The pressure distribution around an aerofoil is usually represented graphically. This can be done in two ways. The first consists of drawing from each point on a sketch of the aerofoil surface a line whose length is proportional to the pressure coefficient at that point, and it is normal to the aerofoil surface. An arrow on each line points outwards for negative p and inwards for positive p. The end of these lines is joined in curves, as illustrated in Figure 4.4.

A more useful presentation, from the analysis point of view, consists of a simple graph in which C_p is plotted against x/c, where x is the distance from the leading edge, measured parallel to the

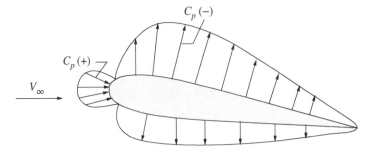

Figure 4.4 Surface pressure distribution over an aerofoil.

---

1 Form drag is the drag due to the pressure distribution around the wing. The drag on the wing due to viscous friction is termed skin-friction drag.

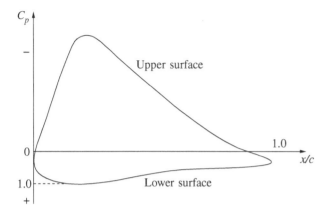

Figure 4.5 Pressure coefficient (C_p) distribution over an aerofoil.

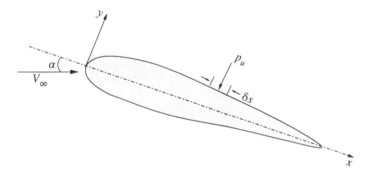

Figure 4.6 An aerofoil at an angle of attack.

chord line, and c is the chord length. By convention, negative values of C_p are plotted above the horizontal axis, and positive values of C_p below the horizontal axis, as shown in Figure 4.5.

From C_p curve, the position of stagnation point ($C_p = 1$) can be read. Also, the total area enclosed by the curve gives the value of the lift coefficient C_L. In a similar manner, graphs can be drawn to give the profile drag coefficient and the pitching moment coefficient. Let us consider an aerofoil shown in Figure 4.6. Let the chord line be on the x-axis and a normal to it through the leading edge be the y-axis. Let X and Y be the components of aerodynamic force in these directions and C_X and C_Y be the coefficients of these forces. Let α be the incidence of the wing. Consider an elemental length δs on the upper surface of the aerofoil. The force per unit span on this element ($p_u \cdot \delta s$) is normal to the surface, where p_u represents the pressure on the upper surface. The component of this force in the y-direction is then ($p_u \cdot \delta x$). Thus, the pressure force on the upper surface contributes

$$-\int_0^c p_u dx$$

to the net force per unit span in the y-direction.

The pressure force on the lower surface contributes

$$\int_0^c p_l dx$$

to the net force per unit span in the y-direction, where p_l is the pressure acting on the lower surface of the aerofoil.

The total force per unit span of the profile in the y-direction is

$$Y = \int_0^c (p_l - p_u)dx$$

The coefficient of this force is

$$C_Y = \frac{Y}{\frac{1}{2}\rho_\infty V_\infty^2\,(c\cdot 1)}$$

$$= \int_0^c \frac{p_l - p_u}{\frac{1}{2}\rho_\infty V_\infty^2\,c}dx$$

This can be expressed as

$$C_Y = \int_0^c \left[\frac{p_l - p_\infty}{\frac{1}{2}\rho_\infty V_\infty^2\,c} - \frac{p_u - p_\infty}{\frac{1}{2}\rho_\infty V_\infty^2\,c} \right]dx$$

$$= \int_0^1 \left(C_{p_l} - C_{p_u} \right)\,d\left(\frac{x}{c}\right)$$

Similarly the force X acting in the x-direction and its coefficient C_X can be obtained by integrating C_p with respect to y/c. For small values of angle of incidence α, C_X is much smaller than C_Y. In terms of C_X and C_Y, we may express the lift and drag coefficients as

$$C_L = C_Y \cos\,\alpha - C_X \sin\,\alpha$$
$$C_D = C_Y \sin\,\alpha + C_X \cos\,\alpha$$

For small values of α, C_L, and C_D simplify to

$$C_L \approx C_Y$$
$$C_D \approx C_Y\cdot\alpha + C_X$$

Thus, the lift coefficient in terms of pressure coefficient is

$$C_L = \int_0^1 \left(C_{p_l} - C_{p_u} \right)\,d\left(\frac{x}{c}\right)$$

which is the area enclosed by the loop in Figure 4.5. The form drag coefficient may be obtained from Figure 4.5 together with the plot of C_p against y/c. Also, the pitching moment coefficient may be expressed as

$$C_M = -\int_0^1 \left(C_{p_l} - C_{p_u} \right)\cdot\frac{x}{c}\cdot d\left(\frac{x}{c}\right)$$

where C_M is the pitching moment coefficient about the leading edge of the aerofoil and may be determined from a graph of C_p against x/c.

4.6 Pressure Distribution Variation with Incidence Angle

The pressure distribution around a two-dimensional aerofoil varies with its incidence to the freestream flow. Pressure distribution around an aerofoil of moderate camber at some representative values of positive and negative angles of incidence is shown in Figure 4.7.

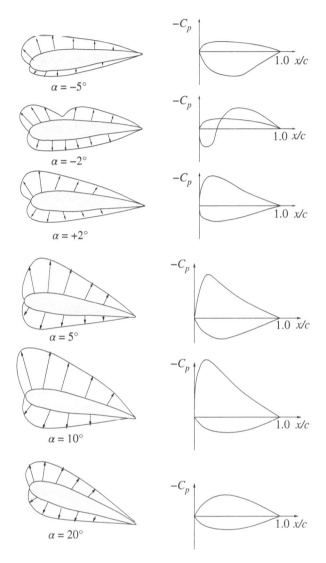

Figure 4.7 Pressure distribution around a cambered aerofoil.

Though qualitative in nature, the plots of C_p versus x/c in Figure 4.7 are good enough for understanding the general nature of the pressure distribution, and the following points may be inferred from these plots:

- At the negative angle of attack of $\alpha = -5°$, the stagnation point is on the upper surface, and the lift is negative.
- At $\alpha = -2°$, for the aerofoil considered, the net area enclosed by the curve of C_p versus x/c is zero, and thus there is a typical zero-lift incidence. The stagnation point is still on the upper surface. The force on the aerofoil reduced to a couple whose moment is nose down, that is, pitching moment at zero lift is negative.
- At the positive angle of attack of $\alpha = 2°$, the stagnation point is on the lower surface. This change occurs at a small positive incidence (also at zero incidence). The lift is positive.

When there is a reasonably large value of lift, it is contributed mainly by the upper surface suction rather than the lower surface pressure. As the incidence increases, the height of the upper surface suction peak increases, and it moves forward. This indicates that the centre of pressure moves forward with increasing incidence.

The zero-lift incidence is negative for a cambered aerofoil and would be zero for a symmetrical aerofoil.

The sudden flattening of the upper surface pressure distribution at high incidence is due to the separation. The effects of this separation quickly spread upstream from the trailing edge and destroy the suction peak. This causes a loss of lift, and further increase of incidence reduces the lift still further. This phenomenon is called *stall*. The lower surface pressure distribution is not immediately affected, because the suction peak is reduced causing the centre of pressure to move aft with the onset of stall.

4.7 The Lift-Curve Slope

Lift curve is a plot of lift coefficient variation with the angle of incidence. A typical lift curve is shown in Figure 4.8. The lift curve essentially shows the way in which the lift coefficient C_L of a two-dimensional aerofoil with moderate positive camber varies with incidence α.

In Figure 4.8, it is seen that the zero-lift angle of attack α_0 is negative; its magnitude in degrees is usually approximately equal to the percentage camber, for example, an aerofoil with 2% camber will have $\alpha_0 \approx -2°$. For symmetrical aerofoil, $\alpha_0 = 0$.

As the angle of attack α increases from α_0, the lift curve is linear over a range of α. However, with increase of incidence, the tendency for flow separation from the upper surface of the aerofoil increases. As the flow begins to separate, the slope of the curve begins to fall off. Around the value of α at which the flow separates, the lift reaches a maximum and begins to decrease. This angle is called the *stalling angle*, and the corresponding value of lift coefficient is denoted by $C_{L_{max}}$. A typical stalling angle would be about 15°, and a typical $C_{L_{max}}$ for a wing without high lift device is about 1.2–1.4.

In the linear region of the lift curve, the lift coefficient is given by

$$C_L = a\left(\alpha - \alpha_0\right)$$

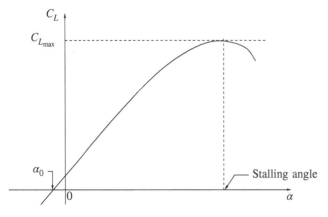

Figure 4.8 Lift curve for a two-dimensional aerofoil.

where a is a constant. Differentiating with respect to α, we have

$$\frac{dC_L}{d\alpha} = a$$

This derivative is called the *lift-curve slope.* The theoretical value of the lift-curve slope is 2π, but the experimental results show that the value of a for a two-dimensional aerofoil is about 5.7 per radian, which is about 0.1 per degree.

In the absence of scale effect, $C_{L_{max}}$ for a given aerofoil has a unique value and always occurs at a fixed stalling angle, irrespective of variations of speed, as in manoeuvres. It is, however, affected by the thickness, camber, and nose radius of the aerofoil and the flow Reynolds number. A low t/c ratio generally implies a fairly low value of $C_{L_{max}}$, increasing with t/c up to 12– 14% but falling off for high values of t/c.

The effect of camber is to increase the circulation round the wing, and hence to increase the lift, at all incidences. However, the increase of camber tends to cause a slightly earlier stall.

A fairly sharp nose (that is, a nose of small radius of curvature) may cause early separation, called *leading-edge stall,* and thus a low value of $C_{L_{max}}$. Note that for sharp nose, the t/c ratio is usually small.

4.8 Profile Drag

The drag of a two-dimensional aerofoil is called *profile drag.* It is the sum of pressure (or form) drag and skin friction drag caused by the viscosity. For a well-designed aerofoil at low α, the wake is thin, and the form drag is significantly smaller than the skin friction drag. As α approaches the stalling value, the separation point moves forward, causing the wake to become thicker. Thickening of wake causes a fairly rapid increase of form drag, and at stall the form drag is greater than the skin friction drag, which does not vary much with α. Thus, over a range of small α, the profile drag coefficient, denoted by C_{D_p}, increases very slowly and is often assumed to be constant. Over a range of α below the stalling angle, the increment in C_{D_p} is approximately proportional to the square of the lift coefficient, and this may be expressed as

$$C_{D_p} = C_{D_0} + kC_L^2$$

where k is a small positive constant and C_{D_0} is the drag coefficient at zero lift.

The lift-to-drag ratio, called *aerodynamic efficiency,* is a measurement of the efficiency of the aerofoil. This may be expressed as

$$\frac{L}{D} = \frac{\frac{1}{2}\rho V^2 S C_L}{\frac{1}{2}\rho V^2 S C_D}$$
$$= \frac{C_L}{C_D}$$

where D is the total drag of the aerofoil, which is the sum of profile drag, induced drag, and so on.

When the profile drag alone is considered, the ratio of C_L/C_D increases with α very rapidly at first, since C_L increases linearly with α, while C_D remains approximately constant. As the stall is approached, C_L increases more slowly, but the C_D increases rapidly. Thus, at a certain α below the stall angle, C_L/C_D reaches a maximum and the then falls for further increase of α. For a two-dimensional aerofoil, the L/D maximum may be of the order of 60–70, or even

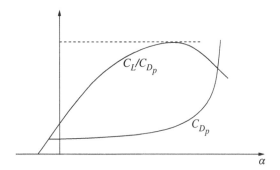

Figure 4.9 Variation of profile drag coefficient and C_L/C_{D_p} with incidence.

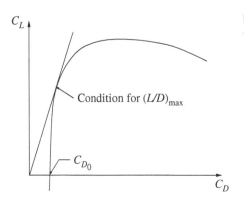

Figure 4.10 Variation of drag coefficient with lift coefficient.

higher, and the α at which this occurs is about 8–10°. Typical variations of C_{D_p} and C_L/C_{D_p} with α for two-dimensional aerofoils are shown in Figure 4.9. At this stage it is essential to note that for three-dimensional aerofoils additional drag occurs, and owing to this increase of drag, L/D maximum becomes smaller. Also, the occurrence of $(L/D)_{max}$ is usually at a much lower α than that for a two-dimensional aerofoil under the same flow condition.

The drag variation for an aerofoil can also be represented as a polar diagram shown in Figure 4.10. It is seen that initially C_L increases more rapidly than C_D, but once C_L begins to decrease (due to the stall), C_D increases fast. The condition for $(L/D)_{max}$ may be found from the drag polar by drawing the tangent to the curve from the origin.

4.9 Pitching Moment

The important aerodynamic forces and moment associated with a flying machine, such as an aircraft, are the lift L, the drag D, and the pitching moment M. The lift and drag forces can be expressed as dimensionless numbers, popularly known as *lift coefficient C_L* and *drag coefficient C_D*, by dividing L and D with $\frac{1}{2}\rho V^2 S$. These forces and moment about some fixed reference point may be determined by experimental methods. The plot of pitching moment coefficient C_M against lift coefficient C_L shows that the relation between C_M and C_L is linear over a considerable range of low incidences, but not as the stall is approached. Thus, relationship between pitching moment and lift coefficient may be expressed as

$$C_M = C_{M_0} + KC_L$$

Figure 4.11 Variation of C_M with C_L for an aerofoil section.

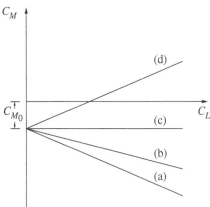

where C_{M_0} is the pitching moment coefficient at zero lift and is a constant and K is a constant whose value depends on the position of the chosen point of reference. At zero lift, $C_M = C_{M_0}$, that is, for zero lift the C_M is a constant and independent of reference point. This is the case where C_M is measured about the aerodynamic centre.

Theoretical curve for C_M against C_L is shown in Figure 4.11. Line (a), for which $dC_M/dC_L \approx -1/4$, is for C_M measured about the leading edge of the aerofoil. Line (b) is for the measurement of C_M about a point between the leading edge and the aerodynamic centre. Line (c), for which the slope is zero, is for the case when C_M is measured about the aerodynamic centre. Line (d) is for the reference point behind the aerodynamic centre. These curves are straight only for moderate values of C_L (or α). As the lift coefficient approached $C_{L_{max}}$, the C_M against C_L curve departs from the straight line. The two possible effects associated with pitching moment and angle of attack, α are the following: (i) Either the pitching moment curve becomes more negative near the stall, thus tending to decrease the incidence, known as a stable break, or (ii) the pitching moment curve may become less negative, leading to increase of incidence and aggravating the stall. Such a characteristic is an unstable break. This type of characteristic is commonly found with highly swept wings.

(iii) The aerodynamic centre, defined as *the point on the aerofoil where the moments are independent of the angle of incidence*, has very important implications. From the definition of aerodynamic centre, it is evident that K is zero so that the pitching moment coefficient is constant at C_{M_0}. The fact that C_M is constant indicates that any increment of lift with increasing α acts through the aerodynamic centre, since they provide no increment in moment about this point. Thus, the position of aerodynamic centre plays a dominant role in the stability of an aircraft. The aerodynamic centre of a low-speed aerofoil is always at or near the quarter-chord point, that is, at a distance of $c/4$ aft of the leading edge.

The concept of aerodynamic centre leads to three methods for representing the forces acting on an aerofoil, as illustrated in Figure 4.12. In Figure 4.12, o is a general reference point, cp is the centre of pressure and ac is the aerodynamic centre. Neglecting the drag, force, and moment shown in Figure 4.12 is identical in Figure 4.3. The advantage of the representation in Figure 4.12c, with ac as the reference point, is that the ac is a fixed point and the pitching moment coefficient about it, C_{M_0}, is a constant. However, it is essential to note that the pitching moment M_0 itself is not necessarily constant; though it does not vary with incidence, it does vary with speed.

With the relative positions of cp, ac, and any other chosen point of reference, say, o, the moment may be obtained as follows.

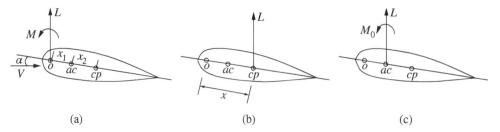

Figure 4.12 Illustration of force and moment at *cp*, *ac*, and a general reference point *o* for an aerofoil section.

Equating the force system of Figure 4.12a,c and taking moment about *o* give

$$M = M_0 - L x_1 \cos \alpha$$

where x_1 is the distance between the *ac* and point *o*. For small value of α, $x_1 \cos \alpha \approx x_1$; thus

$$M = M_0 - L x_1$$

This can be expressed as

$$\frac{M}{\frac{1}{2}\rho V^2 S\, c} = \frac{M_0}{\frac{1}{2}\rho V^2 S\, c} - \frac{L}{\frac{1}{2}\rho V^2 S} \cdot \frac{x_1}{c}$$

This gives

$$C_M = C_{M_0} - C_L \cdot \frac{x_1}{c}$$

where C_{M_0} and x_1 are constants. Differentiating the above relation with respect to C_L, we get

$$\frac{x_1}{c} = -\frac{C_M}{C_L} \tag{4.2}$$

It is seen that the slope of C_M versus C_L curve gives the distance of the *ac* aft of the fixed point of reference *o*. This implies that when the reference point is ahead of the aerodynamic centre, the slope of C_M versus C_L curve is negative and vice versa.

For the force system in Figure 4.12a,b, taking moment about *o*, we have

$$M = -L x$$

This can be expressed as

$$\frac{M}{\frac{1}{2}\rho V^2 S\, c} = -\frac{L}{\frac{1}{2}\rho V^2 S} \cdot \frac{x}{c}$$

or

$$C_M = -C_L \cdot \frac{x}{c}$$

Therefore,

$$\frac{x}{c} = -\frac{C_M}{C_L} \tag{4.3}$$

This gives the distance of the *cp* aft of the fixed point of reference *o* of moment. In Eq. (4.3), x is not a constant because the centre of pressure is a moving point. Equation (4.3) shows that, since C_M/C_L increases with increasing C_L, x/c decreases, and the *cp* moves forward with increasing incidence.

For the force system in Figure 4.12b,c, it is seen that

$$M_0 = -L\,x_2$$

Thus,

$$\frac{M_0}{\frac{1}{2}\rho V^2 S\,c} = -\frac{L}{\frac{1}{2}\rho V^2 S} \cdot \frac{x_2}{c}$$

that is,

$$C_{M_0} = -C_L \cdot \frac{x_2}{c}$$

or

$$\frac{x_2}{c} = -\frac{C_{M_0}}{C_L} \tag{4.4}$$

This gives the distance of *cp* aft of *ac*. Since C_{M_0} is constant, the magnitude of C_{M_0}/C_L decreases as C_L increases. For a wing of positive camber, C_{M_0} is negative, so that x_2/c decreases, showing that the *cp* moves forward as incidence increases. For positive lift, the centre of pressure is always aft of the aerodynamic centre, that is, $x_2/c > 0$; for negative lift, the centre of pressure is ahead of the aerodynamic centre, that is, $x_2/c < 0$. For a wing with negative camber, the converse is true. For a symmetrical wing, $C_{M_0} = 0$, so that the aerodynamic centre and the centre of pressure coincide. Thus for such a wing, the *cp* is fixed. It is essential to note that this applies only at α well below the stalling angle.

4.10   Movement of Centre of Pressure

At low values of incidence, the centre of pressure *cp* moves forward with increasing incidence. This is due to the increase of suction (that is, increase in the negative magnitude of pressure coefficient) and the forward movement of its position on the upper surface of the aerofoil, with increase of incidence angle α. Let us examine the force acting on an aerofoil illustrated in Figure 4.13. At incidence α, the lift L acts through the centre of pressure cp_1. When α is increased to $(\alpha + \Delta\alpha)$, the increase of incidence causes an increment of lift ΔL, and the lift becomes $(L + \Delta L)$. In addition, the increased lift acts at cp_2, which is ahead of cp_1. As shown in Figure 4.13, the new centre of pressure cp_2 lies between the aerodynamic centre *ac* and cp_1. That is, the centre of pressure at higher incidence is forward of the *cp* at a low incidence.

The suction over the upper surface of the aerofoil increases with increase of α, till α reaches α_{stall}. As the stall approaches, the suction peak is reduced. The *cp* begins to move backwards for further increase of α beyond the stalling angle. For an aerofoil with a positive camber, the movement of *cp*

Figure 4.13   Movement of centre of pressure with increase of incidence.

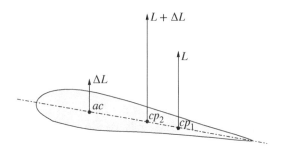

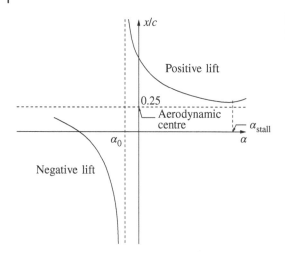

Figure 4.14 Movement of *cp* with incidence for positive and negative lifts.

with varying incidence is illustrated in Figure 4.14. In this plot, *x* is the distance of the *cp* from the leading edge of the aerofoil.

4.11 Finite or Three-Dimensional Wing

For a two-dimensional wing, the span of the wing is assumed to be infinite. Because of the infinite span, the flow past the wing can comfortably be treated as a two-dimensional flow with no velocity component along the wing span. However, for a finite wing, the span is finite, and the flow at the wing tips can easily establish a crossflow, moving from the higher pressure to lower pressure. For a wing experiencing positive lift, the pressure over the lower surface is higher than the pressure over the upper surface. This would cause a flow communication from the bottom to the top, at the wing tips. This tip communication would establish a span-wise variations in the flow. To understand the physics of this flow process, it is necessary to examine the geometrical parameters of the wing.

4.12 Geometrical Parameters of a Finite Wing

The geometrical parameters of a finite or three-dimensional wing are defined in the same way as those for a two-dimensional wing.

The geometrical section of a wing obtained by cutting it by a vertical plane parallel to the centre-line of the aircraft is called *aerofoil section*. The lift generated and the stall characteristics of a wing strongly depend on the geometry of the aerofoil sections that make up the wing. The geometric parameters that dictate the aerodynamic characteristics of the aerofoil section are the leading-edge radius, the mean camber line, the maximum thickness and the thickness distribution of the profile, and the trailing-edge angle.

4.12.1 Leading-Edge Radius and Chord Line

The *chord line* is defined as the shortest (straight) line connecting the leading and trailing edges. The leading edge of aerofoils used in subsonic applications is rounded, with a radius of about 1% of the chord length. The leading edge of an aerofoil is the radius of a circle with its centre on a line tangential to the leading-edge camber connecting tangency points of the upper and lower surfaces with the leading edge. The magnitude of the leading-edge radius has significant effect on the stall characteristics of the aerofoil section.

The *geometrical angle of attack* α is the angle between the chord line and the direction of the undisturbed freestream.

4.12.2 Mean Camber Line

Mean camber line is the locus of the points midway between the upper and lower surfaces of the aerofoil. In other words, mean camber line is the bisector of the aerofoil thickness. The shape of the mean camber line plays an important role in the determination of the aerodynamic characteristics of the aerofoil section. One of the primary effects of camber is to change the zero-lift angle of attack, α_0. For symmetrical aerofoils, zero lift is at $\alpha = 0$, and for cambered aerofoils, zero lift is at negative α for positive camber and vice versa.

The camber has a beneficial effect on the maximum value of the section lift coefficient. If the maximum lift coefficient is high, the stall speed will be low, all other factors being the same. However, it is essential to note that the high thickness and camber necessary for high maximum value of section lift coefficient produce low critical Mach numbers and high twisting moments at high speeds.

4.12.3 Thickness Distribution

The thickness distribution and the maximum thickness strongly influence the aerodynamic characteristics of the aerofoil section. The maximum local velocity to which a fluid element accelerates as it flows around an aerofoil increases as the maximum thickness increases, in accordance with the area–velocity relation for subsonic flow. Thus the value of the minimum pressure is the smallest for the thickest aerofoil. As a result, the adverse pressure gradient associated with the deceleration of the flow, from the location of this pressure minimum to the trailing edge, is greatest for the thick aerofoil. As the adverse pressure gradient becomes larger, the boundary layer becomes thicker. This thickening of boundary layer is likely to cause flow separation, leading to large increase of form drag. Thus, the beneficial effects of increasing the maximum thickness are limited.

For a thin aerofoil section, with relatively small leading-edge radius, boundary layer separation occurs early, not far from the leading edge of the upper (leeward) surface. Because of this, the maximum section lift coefficient for a thin aerofoil section is relatively small. The maximum section lift coefficient increases as the thickness ratio increases.

The thickness distribution for an aerofoil affects the pressure distribution and the character of the boundary layer. As the location of the maximum thickness moves aft, the velocity gradient in the mid-chord region decreases. The favourable pressure gradient associated with this decrease of velocity gradient in the mid-chord region promotes the boundary layer stability and increases the possibility of boundary layer remaining laminar. As we know, the skin friction drag associated with laminar boundary layer is less than that caused by turbulent boundary layer. Further thicker aerofoils benefit more from the use of high-lift devices but have a lower critical Mach number.

4.12.4 Trailing-Edge Angle

The trailing-edge angle influences the location of the *aerodynamic centre*, the point about which the section moment coefficient is independent of angle of attack, α. The aerodynamic centre of this aerofoil section in a subsonic flow is theoretically located at the quarter-chord point.

4.13 Wing Geometrical Parameters

Aircraft wings are made up of aerofoil sections, placed along the span. In an aircraft, the geometry of the horizontal and vertical tails, high-lifting devices such as flaps on the wings and tails, and control surfaces such as ailerons are also made by placing the aerofoil sections in span-wise combinations.

The relevant parameters used to define the aerodynamic characteristics of a wing of rectangular, unswept trapezoidal, swept, and delta configurations are illustrated in Figure 4.15.

4.13.1 Wing Area S

This is the plan surface area of the wing. Thus, the representative area of the wing may be regarded as the product of the span ($2b$) and the average chord ($\bar{c}$). Although a portion of the area may be covered by fuselage, the pressure distribution over the fuselage surface is accounted in the representative wing area.

The wing area that includes the portion of the wing that is effectively cut out to make room for the fuselage is called *gross wing area*, and the wing area that does not include the fuselage is called the *net area*, as illustrated in Figure 4.16.

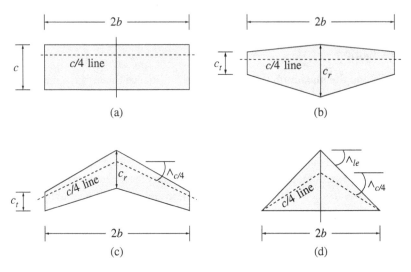

Figure 4.15 Geometric parameters of some wing planforms: (a) rectangular wing, (b) unswept, trapezoidal wing, (c) swept wing, (d) delta wing.

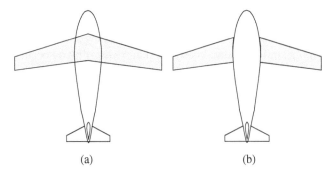

Figure 4.16 (a) Gross and (b) net areas of a wing.

4.13.2 Wing Span 2*b*

This is the distance between the tips of port (left) and starboard (right) wings.

4.13.3 Average Chord $\bar{c}$

This is the geometric average of the chord distribution over the length of the wing span.

4.13.4 Aspect Ratio $\mathcal{R}$

Aspect ratio is the ratio of the span and the average chord. For a rectangular wing, the aspect ratio is

$$\mathcal{R} = \frac{2b}{c}$$

For a non-rectangular wing,

$$\mathcal{R} = \frac{(2b)^2}{S}$$

The $\mathcal{R}$ is a fineness ratio of the wing, and it varies from 35 for sailplanes to about 2 for supersonic fighter planes.

4.13.5 Root Chord c_r

Root chord is the chord at the wing centreline, that is, at the middle of the span, as shown in Figure 4.17. The *tip chord* c_t is the chord at the wing tip.

4.13.6 Taper Ratio λ

Taper ratio is the ratio of the tip chord to root chord, for the wing planforms with straight leading and trailing edges:

$$\lambda = \frac{c_t}{c_r}$$

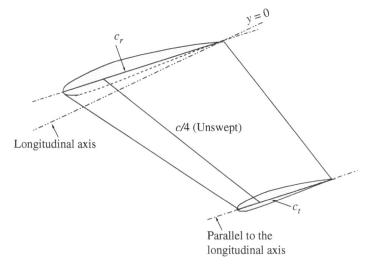

Figure 4.17 Unswept, tapered wing with geometric twist (washout).

The taper ratio affects the lift distribution of the wing. A rectangular wing has a taper ratio of 1.0, while the pointed tip delta wing has a taper ratio of 0.0.

A *tapered wing* is that with tip chord less than the root chord.

4.13.7 Sweep Angle ∧

Sweep angle is usually measured as the angle between the line of 25% chord and a perpendicular to the root chord. The sweep of a wing affects the changes in maximum lift, the stall characteristics and the effects of compressibility.

Sweepback is a feature in which the lines of reference such as the leading and trailing edges of the wing are not normal to the flow direction, and the tip is aft of the root. A wing with sweepback is also referred to as *swept wing*.

A *straight wing* is that which is unswept. If a straight wing is tapered, then the leading edge will necessarily be swept to some extent. The reference line for the purpose of measuring sweepback is usually taken to be the line joining the aerodynamic centre of the wing sections. The angle of sweep is the angle between the appropriate reference line and the normal to the flow direction in the plane of the wing.

Sweepforward is a feature in which the tip of the wing is forward than the root. It is to be noted that only sweepback wings are commonly used and sweep-forward is rare.

4.13.8 Mean Aerodynamic Chord *mac*

Mean aerodynamic chord is an average chord that, when multiplied by the product of the average section moment coefficient, the dynamic pressure, and the wing area, gives the moment for the entire wing. The mean aerodynamic chord is given by

$$mac = \frac{1}{S} \int_{-b}^{+b} [c(y)]^2 dy$$

4.13.9 Dihedral Angle

Dihedral angle is the angle between a horizontal plane containing the root chord and a plane midway between the upper and lower surfaces of the wing. If the wing lies below the horizontal plane, it is termed as *anhedral angle*. The dihedral angle affects the lateral stability of the aircraft.

4.13.10 Geometric Twist

Geometric twist defines the situation where the chord lines for the span-wise distribution of all the aerofoil sections do not lie in the same plane. Thus, there is a span-wise variation in the geometric angle of incidence for the sections. The chord of the root section of the wing shown in Figure 4.17 is inclined at a specified angle relative to the vehicle axis.

The chord at the wing tip, however, is parallel to the longitudinal axis of the vehicle. In this case, where the incidence of the aerofoil sections relative to the vehicle axis decreases toward the tip, the wing has a 'washout'. The wings of most subsonic aircraft have washout to control the span-wise lift distribution and, hence, the boundary layer characteristics. If the angle of incidence increases toward wing tip, the wing has 'washin'.

4.14   Span-Wise Flow Variation

For a lift generating wing, the pressure on the lower surface is higher than the pressure on the upper surface. For a finite wing there will be flow communication at the wing tips, owing to this difference in pressures. This communication at the tips would establish a pressure gradient from the root to tip on the lower surface and from tip to root of the wing on the upper surface, as illustrated in Figure 4.18. Thus there is a span-wise pressure gradient on both surfaces. On the lower surface, pressure decreases from the root to tip, and on the upper surface, pressure increases from the root to tip.

The flow streamlines over the upper surface will be deflected towards the centreline and deflected away from the centreline over the lower surface, as shown in Figure 4.19.

At any given span-wise position, the flow leaving the wing surface at the trailing edge will be moving in two different directions, resulting in continuous shedding of vortices, as shown in Figure 4.20.

However, in practice, this condition is found to be unstable, and these small vortices quickly roll up to form two large vortices positioning near the wing tips, as shown in Figure 4.21.

The tip and trailing vortices cause the flow in the immediate vicinity of the wing, and behind it, to acquire a downward velocity component. This phenomenon is known as *induced downwash* or simply *downwash*. The downwash is usually measured in terms of either the downwash velocity w

Figure 4.18   Span-wise pressure gradient on the wing surfaces.

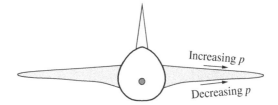

Figure 4.19   Flow pattern on the upper and lower surfaces of a finite wing.

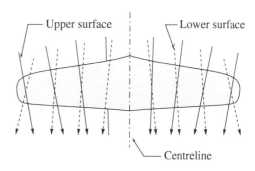

Figure 4.20   Vortex shedding from the trailing edge of a finite wing.

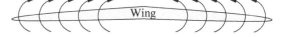

Figure 4.21   Counter rotating vortices at the tips of a finite wing.

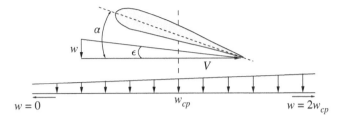

Figure 4.22 Span-wise variation of the strength of the combined bound vortex filaments.

or downwash angle ϵ. The downwash velocity and angle are related, as illustrated by the velocity triangle in Figure 4.22.

The downwash reduces the effective incidence so that for the same lift as the equivalent infinite or two-dimensional aerofoil at incidence α, an incidence of $\alpha = \alpha_\infty + \epsilon$ is required at that section of the aerofoil. Variation of downwash in front of and behind an aerofoil will be as shown in Figure 4.22. As illustrated in Figure 4.22, the downwash will diminish to zero at locations far away from the leading edge and will become almost twice of its magnitude w_{cp} at the centre of pressure, downstream of the trailing edge.

It is seen that

$$\tan \epsilon = \frac{w}{V}$$

where V is the freestream velocity. However, the downwash velocity w is always very small compared with V. Thus, the downwash angle ϵ is a small angle, and $\tan \epsilon \approx \epsilon$. Therefore, we have

$$\boxed{\epsilon = \frac{w}{V}}$$

where ϵ is in radians.

The downwash angle, generally, varies across the span, as shown in Figure 4.22. However, there are some cases in which the downwash is constant across the span, and this is commonly assumed to be approximately true for straight and moderately tapered wings. The downwash also varies in the span-wise direction. It reaches the ultimate value slightly behind the trailing edge; and its mean value at the wing itself can be shown to be one half of this ultimate value. The following are the two important consequences of the downwash:

- The downwash reduces the effective incidence of the wing. This affects both the lift and drag characteristics of the wing adversely.
- The downwash can affect the flow over the tailplane of the aircraft and thus has an important consequence in connection with the stability of the aircraft.

4.15 Lift and Downwash

The lift generated by a wing is due to the pressure distribution over its lower and upper surfaces. A wing made of symmetrical aerofoil sections will experience relatively higher pressure over its lower surface than the upper surface only at a positive incidence. The lift for a symmetrical wing will be zero at zero incidence. However, for a wing made of cambered aerofoil with positive (+ve) camber, lift will be finite even at zero incidence. A negatively (−vely) cambered aerofoil will experience zero lift only at a negative (−ve) incidence. The lift force has its reaction on the downward momentum that is imparted to the air as it flows over the wing. Thus the lift of a wing is equal to the ratio of transport of downward momentum of this air.

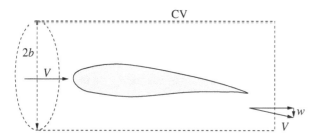

Figure 4.23 A control volume for uniform flow past an aerofoil.

The downward momentum of air passing over a wing is measured in terms of the induced downwash. Consider an aerofoil in a uniform freestream of velocity V, as shown in Figure 4.23. The velocity triangle at the trailing edge shows that the downwash velocity w induces a downward momentum to the flow. Let us consider a cylindrical control volume with diameter equal to the wing span (that is diameter = $2b$).

The area of the cross section of the cylinder of affected air is $\frac{1}{4}\pi(2b)^2$. The rate of mass flow of air influenced by the wing is therefore $\pi\rho b^2 V$. The downwash w imparts a downward momentum to this mass. This rate of transport of downward momentum is therefore $\pi\rho b^2 V w$, and this must be equal to the lift L. This

$$L = \pi\rho b^2 V w$$

$$\frac{1}{2}\rho V^2 S C_L = \pi\rho b^2 V w$$

$$\frac{w}{V} = \frac{C_L}{2\pi}\frac{S}{b^2}$$

However, wing area $S = 2b \times c$, where $2b$ is the span and c is the average chord (here c is used for average chord, instead of $\bar{c}$ for simplicity); therefore,

$$\frac{w}{V} = \frac{C_L}{2\pi}\frac{2b \times c}{b^2}$$

$$= \frac{C_L}{\pi}\frac{c}{b}$$

The ratio $\frac{c}{b} = \frac{2}{\mathcal{R}}$; thus

$$\frac{w}{V} = 2\frac{C_L}{\pi\mathcal{R}} \tag{4.5}$$

This gives the ultimate value of the downwash angle far downstream of the wing. At the wing, the mean downwash is only half of the value, so that it is given by

$$\boxed{\epsilon = \frac{C_L}{\pi\mathcal{R}}} \tag{4.6}$$

In our analysis above, we considered only a finite mass of air affected by a finite wing. If we consider unit span of an infinite wing, the air above this unit span forms a part of a cylinder of infinite radius, and its mass is therefore infinite. Since the lift per unit span is finite, it follows that the downward momentum imparted to this air in unit time is finite, and since the mass of the air is infinite, the induced downwash velocity must be zero. This is natural, since there are no trailing vortices to produce downwash, in the case of a two-dimensional wing. This is evident from Eq. (4.6). For a two-dimensional wing, aspect ratio is infinite, so that ϵ is zero for all values of C_L. Also, it is seen from Eq. (4.6) that the greater the aspect ratio the less the downwash at any given C_L.

4.16   The Lift Curve of a Finite Wing

For a finite wing at a geometric incidence α in an air flow, the effective angle of incidence is reduced by downwash to a value α_{eff}. Because of this, the wing produces only the amount of lift that would be produced by a two-dimensional wing of the same section at an incidence α_{eff}. In other words, a finite wing in general produces a lesser lift than an identical two-dimensional wing placed at the same geometric incidence.

At zero lift there is no downwash; hence there is no reduction of the effective incidence of the finite wing. Thus both two-dimensional and three-dimensional wings have the same zero-lift incidence. The downwash increases with increase of lift, that is, with incidence. Thus the amount of lift lost by a finite wing as compared with a two-dimensional increases with incidence. This implies that both the lift and lift-curve slope of a three-dimensional wing are less than that for a two-dimensional wing at any incidence. The lift curve remains linear over the appropriate incidence range because the downwash ϵ is proportional to lift coefficient C_L.

When the finite wing reaches the geometric incidence at which the two-dimensional wing stalls, its effective incidence is still less than the stalling value. Thus the geometric incidence can continue to increase until the lift coefficient reaches the same maximum value as in the two-dimensional case.

The downwash angle increases as the aspect ratio is reduced. Thus at still lower aspect ratios, the tendencies described above are enhanced, that is, the lift-curve slope is further reduced, and the stalling angle further increased. Typical lift curves for wings as a function of aspect ratio is shown in Figure 4.24.

It is important to note that the curves shown in Figure 4.24 are theoretical curves. In practice, the stall is not delayed as much as the theory suggests, and the maximum lift coefficient consequently falls slightly as aspect ratio is reduced.

The change in lift-curve slope may be found analytically. The geometric incidence and effective angle of incidence are related as

$$\alpha = \alpha_{\text{eff}} + \epsilon$$

where ϵ is the downwash and by Eq. (4.6),

$$\epsilon = \frac{C_L}{\pi \, AR}$$

Therefore,

$$\alpha = \alpha_{\text{eff}} + \frac{C_L}{\pi \, AR}$$

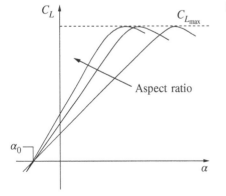

Figure 4.24   Lift curves for wings of different aspect ratio.

Differentiating with respect to C_L, we have

$$\frac{d\alpha}{dC_L} = \frac{d\alpha_{\text{eff}}}{dC_L} + \frac{1}{\pi \mathcal{R}}$$

However, $dC_L/d\alpha = a$ is the lift-curve slope of the finite wing and

$$\frac{dC_L}{d\alpha_{\text{eff}}} = a_{\text{eff}}$$

is the lift-curve slope of a two-dimensional wing of the same section. Thus

$$\frac{1}{a} = \frac{1}{a_{\text{eff}}} + \frac{1}{\pi \mathcal{R}}$$

$$= \frac{a_{\text{eff}} + \pi \mathcal{R}}{\pi \mathcal{R} \, a_{\text{eff}}}$$

or

$$\boxed{a = \frac{\pi \mathcal{R} a_{\text{eff}}}{a_{\text{eff}} + \pi \mathcal{R}}} \tag{4.7}$$

where the lift-curve slope is per radian.

4.17 Induced Drag

The trailing vortices shed from the tips of a finite wing contain energy associated with the rotational velocities. This energy is taken from the airflow over the top and bottom surfaces of the wing, that is, from the freestream velocity. Therefore, some power should be provided to compensate for the energy taken by the tip vortices, to maintain the airflow at a given velocity. This power must be equal to the rate of flow of energy associated with the trailing vortices. This can be regarded as equivalent to an associated drag force on the wing, which should be added to its profile drag. This additional drag is termed *induced drag*. Induced drag can be expressed in terms of the lift coefficient, C_L, and aspect ratio, $\mathcal{R}$, of the wing as follows:

- A finite wing spins the airflow near the tips into what eventually becomes two trailing vortices of considerable core size. The generation of these vortices requires a certain quantity of kinetic energy. The constant expenditure of energy appears to the aerofoil as the *trailing vortex drag*, also known as the *induced drag*.

The freestream flow with velocity V will be deflected downwards due to The downwash, w, establishing a resultant velocity, V_R, as shown in Figure 4.25.
It is seen that

$$V_R^2 = V^2 + w^2$$

The mass flow rate past the wing of span $2b$ (considering a control volume of diameter $2b$) is

$$\dot{m} = \pi \rho V b^2$$

Therefore, the rate of increase of kinetic energy flowing past the wing, due to the downwash, is

$$\frac{1}{2}\dot{m}w^2 = \frac{1}{2}\pi \rho V b^2 w^2$$

Nevertheless, $w^2 = V_R^2 - V^2$; therefore,

$$\frac{1}{2}\dot{m}w^2 = \frac{1}{2}\pi \rho V b^2 (V_R^2 - V^2)$$

Figure 4.25 The forward speed (V) of aerofoil and the resultant velocity (V_R) at the aerofoil.

This must be equal to the power required to overcome the induced drag, D_v. Therefore, the power required is

$$D_v \times V = \frac{1}{2}\pi\rho V b^2 (V_R^2 - V^2)$$

However,

$$D_v = \frac{1}{2}\rho V^2 S C_{D_v}$$

where S is the wing area (span × chord) and C_{D_v} is the induced drag coefficient. Thus

$$\frac{1}{2}\rho V^2 S C_{D_v} V = \frac{1}{2}\pi\rho V b^2 (V_R^2 - V^2)$$

This gives the induced drag coefficient as

$$C_{D_v} = \pi \frac{b^2}{S}\left(\frac{w}{V}\right)^2$$

Consequently, $w/V = \epsilon$; therefore,

$$C_{D_v} = \pi \frac{b^2}{S}\,\epsilon^2$$

Wing area $S = 2b \times c$, and by Eq. (4.5) downwash angle is

$$\epsilon = \frac{2C_L}{\pi \mathcal{R}}$$

Thus,

$$C_{D_v} = \pi \frac{b^2}{2b \times c}\left(\frac{2C_L}{\pi \mathcal{R}}\right)^2$$

The group $\frac{b^2}{2b \times c}$ can be expressed as

$$\frac{2^2 b^2}{2bc}\frac{1}{4} = \frac{2b}{c}\frac{1}{4}$$
$$= \frac{\mathcal{R}}{4}$$

Therefore,

$$C_{D_v} = \frac{\pi \mathcal{R}}{4}\left(\frac{4C_L^2}{\pi^2 \mathcal{R}^2}\right)$$

or

$$\boxed{C_{D_v} = \frac{C_L^2}{\pi \mathcal{R}}} \tag{4.8}$$

This result of induced drag coefficient can also be obtained as follows. Consider the aerofoil in Figure 4.26, showing the two velocity components of the relative wind superimposed on the circulation generated by the aerofoil. In Figure 4.26, L_∞ is the two-dimensional lift, V_R is the resultant velocity, and V is the freestream velocity. Note that the two-dimensional lift is normal to V_R, and the actual lift L is normal to V. The two-dimensional lift is resolved into the aerodynamic forces L and

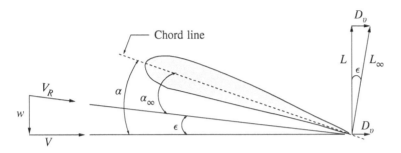

Figure 4.26 Lift and drag caused by the downwash around an aerofoil.

D_v, respectively, normal and against the direction of the velocity V of the aerofoil. Thus, an important consequence of the downwash is the generation of drag D_v. Also, as illustrated in Figure 4.26, the vortex system inducing downwash w tilts the aerofoil in the nose-up direction. In Figure 4.26, V is the forward speed of aerofoil, V_R is the resultant velocity at the aerofoil, α is the incidence, $\epsilon\ (= w/V)$ is the downwash angle, $\alpha_\infty = (\alpha - \epsilon)$, the equivalent two-dimensional incidence, and D_v is the *trailing vortex drag*. The trailing vortex drag is also referred to as *vortex drag* or *induced drag*.

The mean downwash angle at the wing is given by

$$\epsilon = \frac{C_L}{\pi \mathcal{R}}$$

Also,

$$L = L_\infty \cos\ \epsilon$$
$$D_v = L_\infty \sin\ \epsilon$$

Thus

$$D_v = \frac{L}{\cos\ \epsilon}\ \sin\ \epsilon$$
$$= L\ \tan\ \epsilon$$

However, ϵ is small, and thus $\tan\ \epsilon \approx \epsilon$. Hence,

$$D_v = L\ \epsilon$$
$$\frac{1}{2}\rho V^2 S C_{D_v} = \frac{1}{2}\rho V^2 S C_L \epsilon$$
$$C_{D_v} = C_L \epsilon$$
$$= C_L \left(\frac{C_L}{\pi \mathcal{R}}\right)$$

or

$$C_{D_v} = \frac{C_L^2}{\pi \mathcal{R}}$$

An exact analysis present in *Theoretical Aerodynamics*, Ethirajan Rathakrishnan [36] shows that the expression for induced drag coefficient, C_{D_v}, is a function of the span-wise load distribution. The above expression for C_{D_v} can be shown to be the best possible value, that is, the one that gives the smallest induced drag. The expression for C_{D_v} in Eq. (4.8) is obtained by assuming that the downwash angle is constant across the span, and this is true for a wing with elliptical loading. For a wing of elliptical loading, the graph of local section lift coefficient against span-wise position (denoted by distance y from the centreline) is a semi-ellipse, as shown in Figure 4.27.

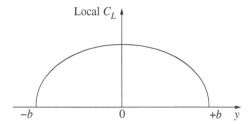

Local C_L

$-b$ 0 $+b$ y

Figure 4.27 Elliptical load distribution over a wing.

For any loading other than elliptical loading, the induced drag is given by

$$C_{D_v} = \frac{k\, C_L^2}{\pi \mathcal{R}} \tag{4.9}$$

where k is a constant for a given wing and it is greater than unity. This constant k is called the *induced drag factor*. The relation is sometimes written as

$$C_{D_v} = \frac{C_L^2}{\pi e \mathcal{R}}$$

where $e = 1/k$, is less than unity, and is called the *Oswald wing efficiency*.

The load distribution is a function of the planform of the wing. An untwisted elliptical wing of constant section has elliptical loading, and this is the optimum shape as far as induced drag is concerned. A straight tapered wing of taper ratio of $2:1$ behaves very much as an elliptic wing. Such wings have e approximately equal to unity. For all other wings, whose loading is not elliptic, $e < 1$. For conventional low-speed wings, value of e is usually about 1.1–1.3.

Equation (4.9) indicates the following:

- There is no induced drag for a wing of infinite aspect ratio (that is, $\mathcal{R} \to \infty$), termed two-dimensional wing.
- The higher the aspect ratio, the less the induced drag, other things being equal.
- There is no induced drag for a wing at zero lift.

4.18 The Total Drag of a Wing

The total drag of a wing is the sum of profile drag and induced drag. If C_D is the total drag coefficient, then

$$C_D = C_{D_0} + C_{D_v}$$

where C_{D_0} is the profile drag coefficient and C_{D_v} is the induced drag coefficient. Also,

$$C_{D_v} = \frac{C_L^2}{\pi e \mathcal{R}}$$

Thus,

$$C_D = C_{D_0} + \frac{C_L^2}{\pi e \mathcal{R}} \tag{4.10}$$

We saw that the profile drag (C_{D_0}) can be assumed to be a constant over a range of low incidence. Thus at low incidences the relationship between C_D and C_L^2 is linear. A plot of C_D variation with C_L^2 is shown in Figure 4.28.

Figure 4.28   Drag polar for a typical wing at low incidence.

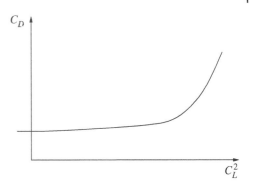

At values of high lift coefficient, the relation is no longer linear, because the profile drag increases rapidly as the stall is approached.

The slope of the linear portion of the curve is $1/(\pi e \mathcal{R})$, and the intercept on the vertical axis is C_{D_0}. Thus for a plot of C_D against C_L^2, the induced drag factor and the profile drag coefficient may be determined.

In fact the profile drag coefficient, C_{D_0}, is not constant, but is more correctly represented by

$$C_{D_0} = C_{D_z} + b' C_L^2$$

Therefore the total drag coefficient becomes

$$C_D = C_{D_z} + b' C_L^2 + \frac{C_L^2}{\pi e \mathcal{R}}$$

or

$$C_D = a + b C_L^2 \tag{4.11}$$

where a and b are constants. The relationship between C_D and C_L^2 is still linear, but the slope of the line will now be greater than $1/(\pi e \mathcal{R})$; in fact, it will be $b' + 1/(\pi e \mathcal{R})$. However, b' is usually small in comparison with $1/(\pi e \mathcal{R})$, and the induced drag factor is generally estimated, with reasonable accuracy, from the drag polar curve.

4.19   Aspect Ratio Effect on Aerodynamic Characteristics

Let us assume that the variations of lift coefficient C_L and drag coefficient C_{D_0} of a two-dimensional wing with incidence α' are known. Then the incidence α and the drag coefficient C_D of a finite wing of the same section and of aspect ratio $\mathcal{R}$, for the same values of C_L, may be found from the following relation, assuming that the wing has elliptic loading, so that the wing efficiency is 1:

$$\alpha = \alpha' + \frac{C_L}{\pi \mathcal{R}}$$
$$C_{D_v} = \frac{C_L^2}{\pi \mathcal{R}}$$
$$C_D = C_{D_0} + C_{D_v}$$

These relations may then be used to determine the characteristics of the finite wing from those of the two-dimensional wing.

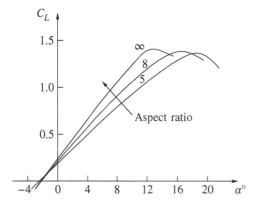

Figure 4.29 Variation of C_L with α and $\mathcal{R}$.

A similar procedure may be used to determine the effects of a change from one finite wing to another. Let subscripts 1 and 2 refer to two finite wings of elliptical loading. For these two wings, we can express

$$\alpha_1 = \alpha' + \frac{C_L}{\pi \mathcal{R}_1}$$

$$\alpha_2 = \alpha' + \frac{C_L}{\pi \mathcal{R}_2}$$

Combining these two relations, we have

$$\alpha_2 = \alpha_1 + C_L \left(\frac{1}{\pi \mathcal{R}_2} - \frac{1}{\pi \mathcal{R}_1} \right) \tag{4.12}$$

Similarly,

$$C_{D_1} = C_{D_0} + \frac{C_L^2}{\pi \mathcal{R}_1}$$

$$C_{D_2} = C_{D_0} + \frac{C_L^2}{\pi \mathcal{R}_2}$$

Combining these two relations, we have

$$C_{D_2} = C_{D_1} + C_L^2 \left(\frac{1}{\pi \mathcal{R}_2} - \frac{1}{\pi \mathcal{R}_1} \right) \tag{4.13}$$

Typical plots of C_L versus α for different aspect ratio is shown in Figure 4.29.

It is seen that the aspect ratio of the wing has a strong influence on the lift-curve slope, $dC_L/d\alpha$. Increase in $\mathcal{R}$ results in increase of lift-curve slope but reduces the stalling angle (that is, $C_{L_{max}}$). It is interesting to note that the zero-lift angle of attack (that is α for $C_L = 0$) is unaffected by the change in $\mathcal{R}$. Lift coefficient variation with drag coefficient for different aspect ratios are shown in Figure 4.30. This plot of C_D variation with C_L is known as *drag polar*.

From this plot, it is seen that the decrease of $\mathcal{R}$ results in increase of C_D and decrease of C_L. The maximum value of lift coefficient is unaffected by aspect ratio variation. However, the lift-to-drag ratio (C_L/C_D) termed *aerodynamic efficiency* is strongly affected by $\mathcal{R}$. A typical variation of C_L/C_D with angle of attack for different aspect ratios is shown in Figure 4.31.

Figure 4.30 Variation of C_L with C_D for different aspect ratio.

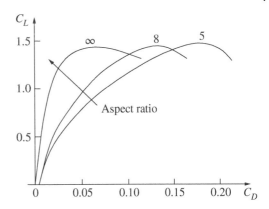

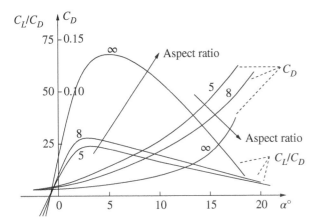

Figure 4.31 Variation of aerodynamic efficiency with α.

4.20 Pitching Moment

As far as pitching moment variations are concerned, there is no significant difference between the general behaviour of a finite wing and a two-dimensional wing. The concept of aerodynamic centre and centre of pressure remains valid, though their positions may be affected by the planform of the wing. For calculating the pitching moment for such a wing, the value of the chord used is a mean chord, such as the geometric mean chord, that is, wing area divided by the wing span.

4.21 The Complete Aircraft

For the complete aircraft, the aerodynamic forces will, in general, be greater than those for the wing alone. Especially the drag will be greater than the drag for the wing alone. However, the lift and drag will generally vary almost the same way as they vary for the wing alone. The definitions of force and moment coefficients remain valid, and the area, S, on which they are based will continue to be the plan area of the wing.

4.22 Straight and Level Flight

In level flight, the lift of the aircraft must be equal to the weight. Thus

$$L = W$$
$$\frac{1}{2}\rho V^2 S C_L = W$$

where ρ is the air density and V is the true airspeed, that is, the actual speed of the aircraft relative to the freestream flow.

The dynamic pressure can be expressed as

$$\frac{1}{2}\rho V^2 = \frac{1}{2}\rho_0 V_e^2$$

where ρ_0 is the sea level density and V_e is called the *equivalent airspeed*. Therefore, in terms of V_e, the lift becomes

$$L = \frac{1}{2}\rho_0 V_e^2 S C_L$$

Thus, the lift coefficient becomes

$$C_L = \frac{W}{\frac{1}{2}\rho V^2 S}$$
$$= \frac{W}{\frac{1}{2}\rho_0 V_e^2 S}$$

It is seen that an increase in speed implies a reduction in lift coefficient.

Since

$$V_e = \sqrt{\frac{2W}{\rho_0 S C_L}} \tag{4.14}$$

it follows that, for a given weight and altitude (that is C_L), the equivalent airspeed of a given aircraft is fixed, that is, V_e does not vary with altitude.

The minimum speed for level flight (that is, the stalling speed) is given by the $C_{L_{\max}}$.
The equivalent stalling speed, V_{es}, is given by

$$V_{es} = \sqrt{\frac{2W}{\rho S C_{L_{\max}}}} \tag{4.15}$$

For a given aircraft at a given weight, V_{es} is fixed, that is, the aircraft will stall at the same equivalent airspeed at all altitudes. Thus, the true stalling speed increases with altitude. This is, of course, the stalling speed in straight and level flight. In a manoeuvre, the lift may be different from the weight, and the equivalent stalling speed may therefore be different. The aircraft always stalls at a given incidence, and therefore, at a given lift coefficient, but not at a given speed.

The wing loading, w, is defined as the weight per unit wing area, that is,

$$\boxed{w = \frac{W}{S}} \tag{4.16}$$

In terms of wing loading, the equivalent airspeed, V_e, and equivalent stalling speed, V_{es}, become

$$V_e = \sqrt{\frac{2w}{\rho C_L}}$$

$$V_{es} = \sqrt{\frac{2w}{\rho C_{L_{max}}}}$$

From this expression for V_{es}, it is evident that it is the wing loading, rather than simply weight, that determines the stalling speed. Also, it is seen that for achieving low stalling speed, the low wing loading must be used.

The load factor, n, for an aircraft is defined as the ratio of the lift, L, to the weight, W:

$$\boxed{n = \frac{L}{W}} \tag{4.17}$$

In straight and level flight, $L = W$, so that $n = 1$. In manoeuvres, $L = nW$, and it is then easily seen that, other things being equal, the stalling speed in the manoeuvre is proportional to $\sqrt{n}$.

Example 4.1
Determine the lift coefficient of an aircraft of mass 50 000 kg, wing area 120 m^2 cruising at 240 m/s, at an altitude where the air density is 0.5 kg/m^3.

Solution
Given: $m = 50\,000$ kg, $S = 120$ m^2, $V = 240$ m/s, and $\rho = 0.5$ kg/m^3.

In cruise, the weight is balanced by the lift. Thus

$$L = m\,g$$

Therefore,

$$\frac{1}{2}\,\rho\,V^2\,S\,C_L = m\,g$$

$$C_L = \frac{2\,m\,g}{\rho\,V^2\,S}$$

$$= \frac{2 \times 50\,000 \times 9.81}{0.5 \times 240^2 \times 120}$$

$$= \boxed{0.284}$$

Example 4.2
An aircraft weighing 20 000 kg flies at 210 m/s at an altitude where the density is 0.8 kg/m^3. The wing span is 12 m and the average chord is 1.6 m. Find the equivalent airspeed and the lift coefficient.

Solution
Given: $m = 20\,000$ kg, $V = 210$ m/s, $2b = 12$ m, $\rho = 0.8$ kg/m^3, and $c = 1.6$ m.

The dynamic pressure is

$$\frac{1}{2}\rho V^2 = \frac{1}{2}\rho_0 V_e^2$$

where ρ, V are the local density and true airspeed and ρ_0, V_e are the sea level density and equivalent airspeed. Therefore,

$$V_e = \sqrt{\rho/\rho_0}\ V$$
$$= \sqrt{0.8/1.225} \times 210$$
$$= \boxed{169.71\ \text{m/s}}$$

In level flight,

$$L = W = m\,g = \frac{1}{2}\rho V^2 S C_L = \frac{1}{2}\rho_0 V_e{}^2 S C_L$$

Therefore,

$$C_L = \frac{m\,g}{\frac{1}{2}\rho_0 V_e{}^2 \times (2b \times c)}$$
$$= \frac{20\ 000 \times 9.81}{\frac{1}{2} \times 1.225 \times 169.71^2 \times (12 \times 1.6)}$$
$$= \boxed{0.58}$$

4.23 Total Drag

The total drag for an aircraft wing is given by Eq. (4.11) is

$$C_D = a + bC_L^2$$

However, for a complete aircraft, the drag due to the other components, such as the fuselage, engine nacelles, tail unit, etc., contribute an additional amount of drag. Since these components are '*non-lifting*', that is, they do not contribute significantly to the lift of the aircraft, their drag is called *parasite drag*. The parasite drag coefficient is usually assumed to be constant.

At this stage it is interesting to note that the drag due to the wing alone is termed *profile drag* and the drag due to the rest of the parts of the aircraft is termed *parasite drag*. This may give an impression that the wing alone is the useful part, and the rest such as the fuselage, engine, tail unit, etc. are considered unwanted. This is because, for an aerodynamicist, generating lift alone is the aim in the design. For an aircraft, the lift required is almost solely generated by the wing alone. Therefore, for what are termed profile or useful part and parasite or unwanted part are from aerodynamicist's point of view. The other extreme to this view is the operator's point of view in which the part of the aircraft that accommodates the payload is the most desirable part. Additionally, if an aircraft can fly only with fuselage, then that would be the most desirable machine. In other words, even the wing would not be of any consequence for such an hypothetical aircraft that can fly only with fuselage.

The parasite drag coefficient has to be added to the constant term of Eq. (4.11). However, the parasite drag coefficient increases with the square of the lift coefficient; therefore there is some increase in the second term bC_L^2 of the C_D expression. The relationship reminds us of the form

$$C_D = a + bC_L^2$$

but the following aspects must be noted.

- The constant a now does not consist only the profile drag coefficient of the wing, but will be higher than that due to the wing alone. In fact, the parasite drag may be considerably higher than the profile drag. However, it is still referred to as the drag coefficient at *zero lift*.

- The variation of C_D with C_L^2 will still be linear over a considerable range. The slope, still given by b, for the whole aircraft will be significantly larger than that due to the induced drag of the wing alone. However, this slope may still be used, as before, to determine what is often called the induced drag factor of the complete aircraft. Typical values for this will be much greater than those quoted above for wing alone.

4.24 Reynolds Number Effect

The factors a and b in the drag coefficient expression are constants only in the absence of scale effect. Increase of Reynolds number tends to reduce separation effects and thus reduces both profile and parasite drag. The value of a is thus reduced, but the effect of the Reynolds number on the value of b is negligible.

The minimum value of the drag coefficient, denoted by $C_{D_{min}}$, is achieved at the theoretical value $C_L = 0$ and is given by $C_{D_{min}} = 0$. It is only a theoretical limit for drag coefficient of the aircraft. In reality an aircraft cannot operate in a straight and level flight with zero-lift coefficient. The actual minimum drag coefficient occurs at the smallest practical value of C_L and thus means at the top speed of the aircraft. Thus increase in Reynolds number will improve this aspect.

The maximum lift coefficient, $C_{L_{max}}$, is an index of the lower limit of the aircraft's speed range; the larger the maximum lift coefficient, the slower the aircraft can fly. Increasing the Reynolds number also will improve this aspect of aircraft's performance.

Thus the ratio of $C_{L_{max}}$ to $C_{D_{min}}$ is often taken as a measure of the overall efficiency of the aircraft, and it improves with increase in Reynolds number. Nevertheless, it is essential to note that the order of magnitude of the changes in Reynolds number that occur simply over the operating range of a given aircraft is generally too small to produce any significant effect. Differences in Reynolds number are very important in interpreting the results of model tested in wind tunnel to the prototype.

4.25 Variation of Drag in Straight and Level Flight

For an aircraft wing with elliptic loading, the induced drag coefficient is given by

$$C_{D_v} = \frac{1}{\pi \mathcal{R}} C_L^2$$

Now, let us consider the aircraft in straight and level flight. Assuming that wing is the only lifting surface, we have

$$L = W = \frac{1}{2}\rho V^2 S C_L$$

where S is the planform area of wing. Therefore,

$$C_L = \frac{W}{\frac{1}{2}\rho V^2 S}$$

Substituting this into C_{D_v}, we have

$$C_{D_v} = \frac{1}{\pi \mathcal{R}}\left(\frac{W}{\frac{1}{2}\rho V^2 S}\right)^2$$

The induced drag is

$$D_v = \frac{1}{2}\rho V^2 S C_{D_v}$$

$$= \frac{1}{2}\rho V^2 S \left[\frac{1}{\pi \mathcal{R}} \left(\frac{W}{\frac{1}{2}\rho V^2 S} \right) \right]^2$$

$$= \frac{1}{\pi \mathcal{R}} \frac{W^2}{\frac{1}{2}\rho V^2 S}$$

The aspect ratio is

$$\mathcal{R} = \frac{\text{Span}}{\text{Chord}}$$

$$= \frac{2b}{c}$$

$$= \frac{2b \times 2b}{c \times 2b}$$

$$= \frac{4b^2}{S}$$

Therefore,

$$D_v = \frac{1}{\pi 4b^2/S} \frac{W^2}{\frac{1}{2}\rho V^2 S}$$

This simplifies to

$$\boxed{D_v = \frac{W^2}{2\pi \rho V^2 b^2}} \tag{4.18}$$

From Eq. (4.18), the following can be observed:

- The induced drag of a wing at any given speed would be low for wings of large span.
- The induced drag is most important at low speed.

Let us analyse the variation of the total drag of the aircraft in straight and level flight, with speed. The drag polar is

$$C_D = a + bC_L^2$$

and the lift coefficient is

$$C_L = \frac{W}{\frac{1}{2}\rho_0 V_e^2 S}$$

where V_e is the equivalent airspeed. Substituting for C_L, we have

$$C_D = a + b \left(\frac{W}{\frac{1}{2}\rho_0 V_e^2 S} \right)^2$$

Thus the drag becomes

$$D = \frac{1}{2}\rho_0 V_e^2 S C_D$$

$$= \frac{1}{2}\rho_0 V_e^2 S \left[a + b \left(\frac{W}{\frac{1}{2}\rho_0 V_e^2 S} \right)^2 \right]$$

$$= \frac{1}{2}\rho_0 V_e^2 Sa + \frac{bW^2}{\frac{1}{2}\rho_0 S} \frac{1}{V_e^2}$$

or

$$D = k_1 V_e^2 + \frac{k_2}{V_e^2} \tag{4.19}$$

where k_1 and k_2 are positive constants. Here, for simplicity, let us refer to the first term on the right-hand side of Eq. (4.19) as the profile drag and to the second term as the induced drag. It is essential to note that these expressions are not strictly correct when dealing with a complete aircraft. Thus the profile drag increases with flight speed, while the induced drag decreases.

It is necessary to know the minimum drag of the wing or aircraft and the speed at which the drag is a minimum. Differentiating Eq. (4.19) with respect to V_e,

$$\frac{dD}{dV_e} = 2k_1 V_e - \frac{2k_2}{V_e^3}$$

Differentiating again with V_e,

$$\frac{d^2D}{dV_e^2} = 2k_1 + \frac{6k_2}{V_e^4}$$

Since $\frac{d^2D}{dV_e^2} > 0$ for all values of V_e, it follows that D is minimum when $\frac{dD}{dV_e} > 0$, that is, when

$$2k_1 V_e - \frac{2k_2}{V_e^3} = 0$$

or

$$\boxed{k_1 V_e^2 = \frac{k_2}{V_e^2}} \tag{4.20}$$

Thus the total drag of the aircraft is minimum when the profile drag and induced drag are equal. The *equivalent minimum drag speed*, V_{md}, from Eq. (4.20) is

$$V_{md}^4 = \frac{k_2}{k_1}$$

$$= \frac{bW^2 / \left(\frac{1}{2}\rho_0 S \right)}{a\frac{1}{2}\rho_0 S}$$

$$= \frac{b}{a} \frac{W^2}{\left(\frac{1}{2}\rho_0 S \right)^2}$$

That is,

$$\boxed{V_{md} = \left(\frac{b}{a} \right)^{1/4} \sqrt{\frac{2W}{\rho_0 S}}} \tag{4.21}$$

From Eq. (4.20),

$$V_{md} = \left(\frac{k_2}{k_1}\right)^{1/4}$$

Substituting this into Eq. (4.19), the minimum drag becomes

$$D_{min} = k_1 V_{md}^2 + \frac{k_2}{V_{md}^2}$$

$$= k_1 \left(\frac{k_2}{k_1}\right)^{1/2} + k_2 \left(\frac{k_1}{k_2}\right)^{1/2}$$

$$= 2 \sqrt{k_1 k_2}$$

$$= 2 \sqrt{a \frac{1}{2}\rho_0 S \times \frac{b\,W^2}{\frac{1}{2}\rho_0 S}}$$

That is,

$$\boxed{D_{min} = 2W \sqrt{ab}} \tag{4.22}$$

From the above results for V_{md} and D_{min}, the following can be observed:

- The equivalent airspeed for minimum drag is dictated only by the weight the aircraft.
- The minimum drag is also independent of everything except weight. That is, D_{min} does not vary with altitude and is directly proportional to the weight.

The minimum drag speed is important due to the following two primary Reasons:

- It represents the optimum condition for the performance of an aircraft in several aspects.
- It also represents the lower limit of the speed range for which the handling qualities are good. Below the minimum drag speed, a slight reduction in speed causes an increase in drag, and hence a further speed reduction; this is an unstable condition requiring continuous correction of the controls by the pilot.

Typical variations of drag coefficient with lift coefficient square, for aircraft in straight and level flight, as a function of wing aspect ratio is illustrated in Figure 4.32. The onset of stall would cause the curves of profile and total drag to deviate, at the lower end of the speed range, from the precise pattern predicted by theory. This is because of the rapid increase in profile drag as the stall is approached.

The fact that the curves curl to the right indicates that, theoretically, at incidences beyond the stall, the lift coefficient has values that are the same as values at incidences before the stall. This implies that the speeds are the same in these two conditions, but the drag is much greater beyond the stall. The curves also show how minimum drag is achieved when the profile drag and induced are equal. Actual curves may deviate from theory at the higher end of the speed range also, because of compressibility effect.

Above results may be obtained by an alternative and more useful method. In straight and level flight, the lift equals the weight, which is a constant, that is, independent of speed or incidence. Therefore L/D is equal to W/D. When the drag is a minimum, the ratio W/D is a maximum. It follows that the minimum drag condition is the same as that for maximum lift-to-drag ratio. Now

$$\frac{L}{D} = \frac{C_L}{C_D} = \frac{C_L}{a + bC_L^2}$$

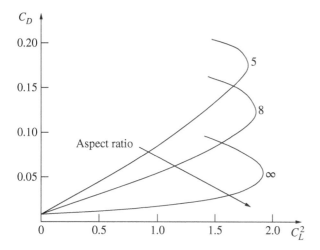

Figure 4.32 C_D variation with C_L^2 as a function of aspect ratio.

Differentiating with respect to C_L, we have

$$\frac{d}{dC_L}\left(\frac{C_L}{a + bC_L^2}\right) = \frac{a + bC_L^2 - C_L 2bC_L}{(a + bC_L^2)^2}$$

$$= \frac{a - bC_L^2}{(a + bC_L^2)^2}$$

This derivative is zero when $a = bC_L^2$, that is,

$$C_L = \sqrt{\frac{a}{b}} \tag{4.23}$$

If

$$C_L > \sqrt{\frac{a}{b}}, \quad \frac{d}{dC_L}\left(\frac{C_L}{a + bC_L^2}\right) < 0$$

If

$$C_L < \sqrt{\frac{a}{b}}, \quad \frac{d}{dC_L}\left(\frac{C_L}{a + bC_L^2}\right) > 0$$

Equation (4.23) is the condition for L/D maximum. Again, since $a = bC_L^2$, we have the profile drag equal to induced drag. Also

$$C_D = a + bC_L^2 = a + a$$

That is, at minimum drag,

$$C_D = 2a \tag{4.24}$$

$$\left(\frac{L}{D}\right)_{max} = \left(\frac{C_L}{C_D}\right)_{max} = \sqrt{\frac{a}{b}}\frac{1}{2a}$$

$$= \frac{1}{2\sqrt{ab}}$$

Thus, at the minimum drag condition,

$$\frac{W}{D} = \frac{1}{2\sqrt{ab}}$$

$$D_{min} = 2W\sqrt{ab}$$

In level flight,

$$L = W$$

$$\frac{1}{2}\rho_0 V_e^2 S C_L = W$$

$$V_e = \sqrt{\frac{2W}{\rho_0 S}}\sqrt{\frac{1}{C_L}}$$

and

$$V_{e_{md}} = \left(\frac{b}{a}\right)^{1/4}\left(\frac{2W}{\rho_0 S}\right)^{1/2}$$

At this stage, it is essential to note that, since we are dealing with a complete aircraft, we should refer to the bC_L^2 term, not as the induced drag coefficient but as the coefficient of drag due to lift. Thus at minimum drag the coefficient of drag due to lift is equal to the drag coefficient at zero lift.

Example 4.3
Find the induced drag coefficient of the aircraft of mass of 50 000 kg, span of 35 m and average chord of 3.5 m in level fight at sea level with a speed of 200 m/s.

Solution
Given: $m = 50\ 000$ kg, $2b = 35$ m, $c = 3.5$ m, and $V = 200$ m/s. At sea Level, air density is $\rho = 1.225\ \text{kg/m}^3$.

For an aircraft in level flight,

$$L = W = m\,g$$

Therefore,

$$\frac{1}{2}\rho V^2 S\, C_L = m\,g$$

Thus the lift coefficient becomes

$$C_L = \frac{2\,m\,g}{\rho\,V^2 S}$$

$$= \frac{2\,m\,g}{\rho\,V^2\,(2b \times c)}$$

$$= \frac{2 \times 50\ 000 \times 9.81}{1.225 \times 200^2 \times (35 \times 3.5)}$$

$$= 0.163$$

The aspect ratio is

$$\mathcal{R} = \frac{2b}{c}$$

$$= \frac{35}{3.5}$$

$$= 10$$

The induced drag coefficient is given by

$$C_{D_v} = \frac{1}{\pi \, R} \, C_L^2$$

$$= \frac{1}{\pi \times 10^2} \times 0.163^2$$

$$= 8.46 \times 10^{-5}$$

Example 4.4

Determine the induced drag and its coefficient of an aircraft weighing 100 kN cruising at sea level with a speed of 130 m/s, if the wing span and average chord are 20 m and 1.2 m, respectively.

Solution

Given: $W = 100\ 000$ N, $2b = 20$ m, $c = 1.2$ m, $\rho = 1.225$ kg/m^3, and $V = 130$ m/s.

In level flight,

$$L = W = \frac{1}{2}\rho V^2 S C_L$$

Therefore, the lift coefficient becomes

$$C_L = \frac{2W}{\rho V^2 S}$$

$$= \frac{2W}{\rho V^2 (2b\ c)}$$

$$= \frac{2 \times 100\ 000}{1.225 \times 130^2 \times (20 \times 1.2)}$$

$$= 0.4025$$

By Eq. (4.18), the induced drag is

$$D_v = \frac{W^2}{2\rho\pi V^2 b^2}$$

$$= \frac{100\ 000^2}{2 \times 1.225 \times \pi \times 130^2 \times 10^2}$$

$$= \boxed{768.77\ \text{N}}$$

The induced drag is given by

$$D_v = \frac{1}{2}\rho V^2 S C_{D_v}$$

Therefore,

$$C_{D_v} = \frac{D_v}{\frac{1}{2}\rho V^2 S}$$

$$= \frac{768.77}{\frac{1}{2} \times 1.225 \times 130^2 \times (20 \times 1.2)}$$

$$= \boxed{3.09 \times 10^{-3}}$$

Example 4.5

An aircraft has a span of 20 m and chord of 1 m, weighing 150 kN cruises at sea level at 120 m/s. If the aerodynamic efficiency is 4 and the Oswald wing efficiency is 0.95, determine the minimum drag speed and the minimum drag. Verify the $L/D_{\min}$ with the relation $1/(2\sqrt{ab})$.

Solution

Given: $2b = 20$ m, $c = 1$ m, $W = 150\ 000$ N, $V = 120$ m/s, $L/D = 4$, and $e = 0.95$.

In level flight,

$$L = W = \frac{1}{2}\rho V^2 S C_L$$

Therefore, the lift coefficient is

$$C_L = \frac{2W}{\rho V^2 S}$$

$$= \frac{2 \times 150\ 000}{1.225 \times 120^2 \times (20 \times 1)}$$

$$= 0.85$$

The aerodynamic efficiency is

$$L/D = C_L/C_D = 4$$

Therefore,

$$C_D = C_L/4 = 0.85/4$$

$$= 0.2125$$

The drag coefficient is given by

$$C_D = a + bC_L{}^2$$

where

$$b = \frac{1}{\pi e \mathcal{R}}$$

$$= \frac{1}{\pi e (2b/c)}$$

$$= \frac{1}{\pi \times 0.95 \times (20/1)}$$

$$= 0.0167$$

Therefore,

$$a = C_D - bC_L{}^2$$

$$= 0.2125 - 0.0167 \times 0.85^2$$

$$= 0.2004$$

By Eq. (4.21), the minimum drag speed is

$$V_{\text{md}} = \left(\frac{b}{a}\right)^{1/4} \left(\frac{2W}{\rho_0 S}\right)^{1/2}$$

$$= \left(\frac{0.0167}{0.2004}\right)^{1/4} \times \left(\frac{2 \times 150\ 000}{1.225 \times (20 \times 1)}\right)^{1/2}$$

$$= \boxed{59.45 \text{ m/s}}$$

By Eq. (4.22), the minimum drag is

$$D_{\text{min}} = 2W\sqrt{ab}$$

$$= (2 \times 150\ 000) \times \sqrt{0.2004 \times 0.0167}$$

$$= \boxed{17\ 355.15 \text{ N}}$$

The minimum drag coefficient is

$$C_{D\min} = \frac{D_{\min}}{\frac{1}{2}\rho V_{\mathrm{md}}^2 S}$$

$$= \frac{17\ 355.15}{\frac{1}{2} \times 1.225 \times 59.45^2 \times (20 \times 1)}$$

$$= 0.4008$$

Thus,

$$C_L/C_{D\min} = 0.85/0.4008 = 2.121$$

Thus, $(L/D)_{\max}$ becomes

$$(L/D)_{\max} = L/D_{\min}$$

$$= 150\ 000/17\ 355.15$$

$$= 8.643$$

The $(L/D)_{\max}$ is also given by

$$(L/D)_{\max} = \frac{1}{2\sqrt{ab}}$$

$$= \frac{1}{2 \times \sqrt{0.2004 \times 0.0167}}$$

$$= 8.643$$

4.26 The Minimum Power Condition

An important condition associated with aircraft performance is that the power required to overcome drag is a minimum.

Power required can be expressed as

$$P = DV$$

where D is the drag and V is the flight speed. Therefore,

$$P = \left(\frac{1}{2}\rho V^2 SC_D\right) V$$

where the density ρ and wing area S are constants in a steady level flight. This

$$P \propto C_D V^3$$

In level flight,

$$L = W = \frac{1}{2}\rho V^2 SC_L$$

Using this, we get the power as

$$P \propto \frac{C_D}{C_L^{3/2}} \tag{4.25}$$

Thus, the power required is a minimum when $C_L^{3/2}/C_D$ is a maximum. We have the drag coefficient as

$$C_D = a + bC_L^2$$

Therefore,

$$\frac{C_L^{3/2}}{C_D} = \frac{C_L^{3/2}}{a + bC_L^2}$$

Differentiating with respect to C_L,

$$\frac{d}{dC_L}\left(\frac{C_L^{3/2}}{C_D}\right) = \frac{\left(a + bC_L^2\right)\frac{3}{2}C_L^{1/2} - C_L^{3/2}\,2bC_L}{\left(a + bC_L^2\right)^2}$$

For power minimum,

$$\frac{d}{dC_L}\left(\frac{C_L^{3/2}}{C_D}\right) = 0$$

Therefore,

$$\frac{\left(a + bC_L^2\right)\frac{3}{2}C_L^{1/2} - C_L^{3/2}\,2bC_L}{\left(a + bC_L^2\right)^2} = 0$$

$$\frac{3}{2}\left(a + bC_L^2\right)C_L^{1/2} - 2bC_L C_L^{3/2} = 0$$

$$\frac{3}{2}\left(a + bC_L^2\right) - 2bC_L^2 = 0$$

$$3a - bC_L^2 = 0$$

This gives

$$\boxed{C_L = \sqrt{\frac{3a}{b}}} \tag{4.26}$$

For this case of minimum power, the equivalent airspeed becomes

$$\boxed{V_{e_{mp}} = \sqrt{\frac{2W}{\rho_0 S}}\left(\frac{b}{3a}\right)^{1/4}} \tag{4.27}$$

From Eqs. (4.21) and (4.27), we get

$$V_{e_{mp}} = \frac{1}{3^{1/4}}V_{e_{md}}$$

that is,

$$\boxed{V_{e_{mp}} = 0.76\,V_{e_{md}}} \tag{4.28}$$

4.27 Minimum Drag-to-Velocity Ratio

Minimum drag-to-velocity ratio is another important condition of operation. The ratio of drag to velocity is

$$\frac{D}{V} = \frac{\frac{1}{2}\rho V^2 S C_D}{V}$$

In a level flight, the density ρ and wing area S are constants. Therefore,

$$\frac{D}{V} \propto C_D \, V$$

However, in level flight,

$$V \propto \frac{1}{C_L^{1/2}}$$

Therefore,

$$\frac{D}{V} \propto \frac{C_D}{C_L^{1/2}}$$

The drag-to-velocity ratio is a minimum when $C_L^{1/2}/C_D$ is a maximum. Consequently,

$$\frac{C_L^{1/2}}{C_D} = \frac{C_L^{1/2}}{a + bC_L^2}$$

Differentiating with respect to C_L, we have

$$\frac{d}{dC_L} \left(\frac{C_L^{1/2}}{C_D} \right) = \frac{\left(a + bC_L^2\right) \frac{1}{2} C_L^{-1/2} - C_L^{1/2} \, 2bC_L}{\left(a + bC_L^2\right)^2}$$

For D/V minimum, the condition is

$$\frac{d}{dC_L} \left(\frac{C_L^{1/2}}{C_D} \right) = 0$$

Thus,

$$\frac{\left(a + bC_L^2\right) \frac{1}{2} C_L^{-1/2} - C_L^{1/2} \, 2bC_L}{\left(a + bC_L^2\right)^2} = 0$$

$$\frac{1}{2} \left(a + bC_L^2\right) C_L^{-1/2} - C_L^{-1/2} \, 2bC_L = 0$$

$$\frac{1}{2} \left(a + bC_L^2\right) - 2bC_L^2 = 0$$

$$a - 3bC_L^2 = 0$$

that is,

$$\boxed{C_L = \sqrt{\frac{a}{3b}}} \tag{4.29}$$

The speed at which D/V is minimum in level flight is achieved as follows.

In level flight,

$$L = W = \frac{1}{2}\rho_0 V_e^2 S C_L$$

$$V_e^2 = \frac{2W}{\rho_0 S C_L}$$

Substituting for C_L from Eq. (4.29), we have the speed for D/V minimum as

$$\boxed{V_{e_{\min D/V}} = \left(\frac{2W}{\rho_0 S} \right)^{1/2} \left(\frac{3b}{a} \right)^{1/4}} \tag{4.30}$$

From Eqs. (4.21) and (4.30), we get the relationship between $V_{e_{\min D/V}}$ and $V_{e_{md}}$ as

$$\boxed{V_{e_{\min D/V}} = (3)^{1/4}\, V_{e_{md}}}$$ (4.31)

From the above results for minimum drag, minimum power, and minimum D/V the following can be inferred:

- The condition for minimum drag, minimum power, and minimum D/V for a given aircraft always occurs at given values of lift coefficient, that is, at given incidence, independent of both altitude and weight.
- The corresponding speeds are functions of weight, but expressed as equivalent airspeed, which are independent of altitude.
- The minimum power speed is always a given fraction, and the minimum D/V speed a given multiple, of the minimum drag speed.

4.28 The Stall

Stalling of an aircraft is characterised by loss of lift and increase of drag. This is due to the effects of separation. Thus if stalling occurs in flight, the aircraft will lose height, unless some action is taken to prevent it. The aircraft behaviour and handling at and near the stall depend on the design of the wing.

4.28.1 The Effect of Wing Section

The shape of the aerofoil that constitutes the wing section affects the behaviour of the wing near stall. With some aerofoil section, the stall occurs very suddenly, and the decrease of lift is very large. For some aerofoil sections, the approach to the stall is gradual, and the drop in lift is not large. The drag curves shown in Figure 4.33 illustrate these characteristic differences.

In practice, it is desirable that the stall does not occur suddenly and that the pilot should have adequate warning, in terms of handling qualities of the aircraft, of the approach of the stall. This warning generally takes the form of buffeting and general lack of response to the controls. If a particular wing is such that it stalls too suddenly, it may be necessary to provide some artificial pre-stalling warning device.

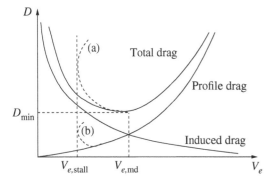

Figure 4.33 Drag curve for a wing for which stall is (a) sudden and (b) gradual.

Features of aerofoil section that affect the behaviour near stall are the Following:

- Radius of curvature of the leading edge.
- Thickness-to-chord ratio (t/c).
- Camber and particularly the amount of camber near the leading edge.
- Chord-wise location of the points of maximum thickness and maximum camber.

Usually sudden stall will occur for wings with sharper nose, thicker profile, and the maximum thickness and maximum camber far away from the leading edge.

4.28.2 Wing Planform Effect

For a finite wing, separation does not necessarily occur simultaneously at all span-wise position. If the wing is rectangular, separation tends to occur first near the wing root and spreads out to the outboard regions of the wing. The consequent reduction in lift is initially felt near the rolling axis of the aircraft, and the aircraft loses height, but in doing so more or less on an even keel. The loss of lift is also felt ahead of the centre of gravity of the aircraft, so that the nose drops, speed increases, and incidence is reduced. Thus, the natural tendency of the aircraft is to move away from the high incidence condition that gives rise to the stall. This is the most desirable kind of response to the stall.

For a highly tapered wing, separation tends to occur first near the wing tips and then spreads inwards. This means that the reduction in lift is first felt at the tips. If one wing stalls before the other, then a considerable moment will be set up about the rolling axis of the aircraft, one wing will drop, and a rolling motion will be initiated. When this happens, a spin is likely to develop unless recovery action is taken by the pilot. Also, since the ailerons, which control the rolling of the aircraft, are located near the wing tips, tip stalling results in the loss of lateral control.

Sweep wings also tend to stall first near the tips. In this case, there is an additional adverse effect. Since the wing tips are situated well aft of the centre of the aircraft, loss of lift at the tips causes the nose to come up rapidly, and this increases the incidence. To get away from the stalled condition, it is necessary to reduce the incidence. However, the reduction in incidence may become disastrous if the flight is near the ground, as in landing or take-off, when the aircraft is operated at high incidence. Even well above the ground, it generally results in the loss of control in pitch. This phenomenon is known as *pitch-up* and is a dangerous characteristic of many high-speed aircraft.

Various measures may be employed to prevent tip stalling. The following are some of the popular measures taken for the purpose:

- *Washout*: If the incidence is less at the tip than that at the root, then stalling at the tip may be delayed until after the stall has occurred at the root. There is, of course, a limit to the amount of washout that can be used. If too much is used, then at high speed, when the overall incidence is low, the incidence at the tip may be less than the zero-lift incidence, so that the tip carries a download. This clearly reduces the efficiency of the wing.
- Variations in section shape may be used across the span of the wing. In particular, different amounts of camber may be used in different span-wise position, since this gives a measure of control over local maximum lift coefficient and stalling angle.
- A part-span leading edge may be used on the outboard sections of the wing. Such slots are described as high-lift devices. They tend to prevent separation of the airflow in the regions behind them. It is used only near the wing tips.
- Certain special planforms may be used. One of the advantages of delta and crescent wings, for instance, is that they reap some of the advantages derived from a considerable amount of sweep-back, without incurring the penalty of tip stalling.

4.29 The Effect of Protuberances

Any protuberances on the wing or elsewhere on the aircraft can significantly affect the stalling pattern by causing local flow separations. The main cause of such effects is the positioning of the engines, since separation may occur near engine nacelles or spring from the intakes of jet engines.

Some protuberances on the wing to control stalling behaviour are the following:

- *Vortex generators*: These are small aerofoils of large camber so placed as to introduce swirling motions in the boundary layer. This energising of the boundary layer tends to prevent separation.
- *Boundary layer fencing*: These are small plates, employed to prevent the outward drift of the boundary layer that is a factor in causing the tip stall. Notches at the leading edge of the wing also tend to produce a similar effect.

4.30 Summary

An aerofoil is a *streamlined body that would experience the maximum aerodynamic efficiency (that is, maximum lift-to-drag ratio) compared with any other body under identical flow conditions.*

The major geometrical parameters of an aerofoil are the following. *Chord line* is defined as the shortest (straight) line connecting the leading and trailing edges. *Chord c* is the shortest distance between the leading and trailing edges. *Maximum thickness* t_{max} is measured normal to the chord line. The t/c is generally about 12–14% for subsonic aerofoils and as low as 3% or 4% for supersonic aerofoils. *Camber line* is the bisector of the profile thickness from the chord of the profile. *Camber* is the maximum deviation d of the *camber line* from the chord.

The attitude of the aerofoil is expressed by the angle between the chord line and the freestream velocity vector. This angle is called the *incidence* or *angle of attack*.

Aerodynamic force acting on an aircraft is the force due to the pressure distribution around it, caused by the motion of the aircraft.

The aerodynamic force acts along a line whose intersection with the chord line is called the *centre of pressure cp* of the aerofoil. The aerodynamic force normal to the freestream flow direction is the *lift*, and the one parallel to the freestream flow direction is the *drag*.

The important aerodynamic forces and moment associated with an aircraft, are the lift L, drag D, and pitching moment M.

The pitching moment is the moment of the aerodynamic force about an axis perpendicular to the plane of symmetry.

Pressure distribution around the aerofoil results in the lift, pitching moment and form drag acting on the aerofoil, and the position of its centre of pressure.

At the stagnation point, C_p has the limiting maximum of 1. However, there is no overall limit for its lower value.

The zero-lift incidence is negative for a cambered aerofoil and would be zero for a symmetrical aerofoil.

A fairly sharp nose (that is, a nose of small radius of curvature) may cause early separation, called *leading-edge stall*, and thus a low value for $C_{L_{max}}$.

The drag of a two-dimensional aerofoil is called *profile drag*. It is the sum of pressure (or form) drag and skin friction drag caused by the viscosity.

The lift-to-drag ratio, called *aerodynamic efficiency*, is a measure of the efficiency of the aerofoil.

The important aerodynamic forces and moment associated with a flying machine are the lift L, the drag D, and the pitching moment M. The lift and drag forces can be expressed as dimensionless numbers, popularly known as *lift coefficient C_L* and *drag coefficient C_D*, by dividing L and D with $\frac{1}{2}\rho V^2 S$.

The relationship between pitching moment and lift coefficient may be expressed as

$$C_M = C_{M_0} + KC_L$$

where C_{M_0} is the pitching moment coefficient at zero lift and is a constant and K is a constant whose value depends on the position of the chosen point of reference.

The aerodynamic centre is *the point on the aerofoil where the moments are independent of the angle of incidence.*

For a two-dimensional wing, the span of the wing is assumed to be infinite.

The geometrical section of a wing obtained by cutting it by a vertical plane parallel to the centre-line of the aircraft is called *aerofoil section.*

The *chord line* is defined as the shortest (straight) line connecting the leading and trailing edges.

The *geometrical angle of attack α* is the angle between the chord line and the direction of the undisturbed freestream.

Mean camber line is the locus of the points midway between the upper and lower surfaces of the aerofoil. In other words, mean camber line is the bisector of the aerofoil thickness.

For a thin aerofoil section, with relatively small leading-edge radius, boundary layer separation occurs early, not far from the leading edge of the upper (leeward) surface. The thickness distribution for an aerofoil affects the pressure distribution and the character of the boundary layer.

The trailing-edge angle influences the location of the *aerodynamic centre*, the point about which the section moment coefficient is independent of angle of attack, α. The aerodynamic centre of this aerofoil section in a subsonic flow is theoretically located the quarter-chord point.

Aircraft wings are made up of aerofoil sections, placed along the span. In an aircraft, the geometry of the horizontal and vertical tails, high-lifting devices such as flaps on the wings and tails, and control surfaces such as ailerons are also made by placing the aerofoil sections in span-wise combinations.

Wing area S is the plan surface area of the wing. Thus, the representative area of the wing may be regarded as the product of the span ($2b$) and the average chord ($\bar{c}$). The wing area that includes the portion of the wing that is effectively Cut out to make room for the fuselage is called *gross wing area*, and the wing area that does not include the fuselage is called the *net area.*

Wing span is the distance between the tips of port and starboard wings.

Average chord is the geometric average of the chord distribution over the length of the wing span.

Aspect ratio is the ratio of the span and the average chord. The $\mathcal{R}$ is a fineness ratio of the wing, and it varies from 35 for sailplanes to about 2 for supersonic fighter planes.

Root chord is the chord at the wing centreline, that is, at the middle of the span. The tip chord is the chord at the wing tip.

Taper ratio is the ratio of the tip chord to root chord.

A *tapered wing* is that with tip chord less than the root chord.

Sweepback is a feature in which the lines of reference such as the leading and trailing edges of the wing are not normal to the flow direction and the tip is aft of the root.

Sweepforward is a feature in which the tip of the wing is forward than the root. It is to be noted that only sweepback wings are commonly used and sweepforward is rare.

Dihedral angle is the angle between a horizontal plane containing the root chord and a plane midway between the upper and lower surfaces of the wing. If the wing lies below the horizontal plane, it is termed as *anhedral angle.*

Geometric twist defines the situation where the chord lines for the span-wise distribution of all the aerofoil sections do not lie in the same plane. In the case, where the incidence of the aerofoil sections relative to the vehicle axis decreases toward the tip, the wing has a 'washout'. The wings of most subsonic aircraft have washout to control the span-wise lift distribution and, hence, the boundary layer characteristics. If the angle of incidence increases toward wing tip, the wing has 'washin'.

The tip and trailing vortices cause the flow in the immediate vicinity of the wing, and behind it, to acquire a downward velocity component. This phenomenon is known as *induced downwash* or simply *downwash*.

The following are the two important consequences of the downwash: First, the downwash reduces the effective incidence of the wing. Second, this can affect the flow over the tailplane of the aircraft.

The lift generated by a wing is due to the pressure distribution over its lower and upper surfaces. The lift force has its reaction in the downward momentum that is imparted to the air as it flows over the wing. Thus the lift of a wing is equal to the ratio of transport of downward momentum of this air.

The trailing vortices shed from the tips of a finite wing contain energy associated with the rotational velocities.

A finite wing spins the airflow near the tips into what eventually becomes two trailing vortices of considerable core size. The generation of these vortices requires a quantity of kinetic energy. The constant expenditure of energy appears to the aerofoil as the *trailing vortex drag*, also known as the *induced drag*.

There is no induced drag for a wing of infinite aspect ratio. The higher the aspect ratio, the less the induced drag, other things being equal. There is no induced drag for a wing at zero lift.

The total drag of a wing is the sum of profile drag and induced drag.

As far as pitching moment variations are concerned, there is no significant difference between the general behaviour of a finite wing and a two-dimensional wing.

For the complete aircraft, the aerodynamic forces will, in general, be greater than those for the wing alone. Especially the drag will be greater than the drag for the wing alone. However, the lift and drag will generally vary almost the same way as they vary for the wing alone. The definitions of force and moment coefficients remain valid, and the area, S, on which they are based will continue to be the plan area of the wing.

In level flight, the lift of the aircraft must be equal to the weight. Thus

$$L = W$$
$$\frac{1}{2}\rho V^2 S C_L = W$$

The dynamic pressure can be expressed as

$$\frac{1}{2}\rho V^2 = \frac{1}{2}\rho_0 V_e^2$$

where ρ_0 is the sea level density and V_e is called the *equivalent airspeed*. Therefore, in terms of V_e, the lift becomes

$$L = \frac{1}{2}\rho_0 V_e^2 S C_L$$

For a given weight and altitude (that is C_L), the equivalent airspeed of a given aircraft is fixed, that is, V_e does not vary with altitude.

The minimum speed for level flight (that is, the stalling speed) is given by the $C_{L_{max}}$.

For a given aircraft at a given weight, V_{es} is fixed, that is, the aircraft will stall at the same equivalent airspeed at all altitudes. Thus, the true stalling speed increases with altitude.

The wing loading, w, is defined as the weight per unit wing area used.

The load factor, n, for an aircraft is defined as the ratio of the lift, L, to the weight, W,

$$\boxed{n = \frac{L}{W}}$$

In straight and level flight, $L = W$, so that $n = 1$. In manoeuvres, $L = nW$, and it is then easily seen that, other things being equal, the stalling speed in the manoeuvre is proportional to $\sqrt{n}$.

Problems

4.1 If the local pressure coefficient on the top surface of a wing is -0.44 while flying at 180 m/s, find the local flow speed at that point.

4.2 A glider of mass 444 kg is cruising at 20 m/s at sea level. If the wing span and chord are 15 and 1.2 m, determine the lift coefficient.

4.3 The aerodynamic efficiency of a glider of mass 510 kg flying level at 30 m/s at sea level is 4. If the span and chord are 15 and 1.1 m, respectively, determine the drag.

4.4 If the lift-curve slope of a wing with zero degree angle of attack as $-3°$ is 4, determine the angle of attack for lift coefficient 0.6.

4.5 If a two-dimensional wing of aspect ratio 15 has a lift coefficient of 0.9 when the aerodynamic efficiency is 25, determine the zero-lift drag coefficient.

4.6 Find the chord of a wing span 22 m and area 26 m^2.

4.7 Find the percentage change in equivalent stalling speed of an aircraft cruising at 5000 m made to cruise at 10 000 m altitude.

4.8 An aircraft weighing 5000 N flies level at 28 m/s at sea level. If the wing span and the chord are 15 and 1.6 m, respectively, and the Oswald wing efficiency is 0.9, determine the induced drag.

4.9 An aircraft weighing 200 kN flies level at minimum power state at an altitude where the density is 0.9 kg/m^3. The flight speed, wing area, aspect ratio, and wing efficiency are 160 m/s, 45 m^2, 30, and 0.8, respectively. Determine the aerodynamic efficiency.

4.10 An aircraft weighing 300 kN flies level at 220 m/s at an altitude where the air density is 0.8 kg/m^3. The area, aspect ratio, and efficiency of the wing are 45 m^2, 38 and 1, respectively. Determine the minimum drag and the equivalent minimum drag speed of the aircraft.

4.11 (i) An aircraft weighing 200 kN flies level at 140 m/s at sea level. The zero-lift drag coefficient of the wing with aspect ratio of 12 is 0.01. (ii) If the aspect ratio of the wing is doubled, determine the percentage change in the aerodynamic efficiency of the wing.

5

High-Lift Devices

5.1 Introduction

During take-off and landing of an aircraft, high values of lift coefficient are required to maintain flight at the desired low speeds. For most low-speed aircraft, the maximum lift coefficient is about 1.3 or 1.4. Thus, the stalling speed will be dictated by these maximum values of lift coefficient. If low stalling speeds are required, high values for the maximum lift coefficient must be achieved.

The lift coefficient of an aerofoil is strongly influenced by its camber. The higher the camber, the higher the lift coefficient. Thus, by simply increasing the camber in the design of the aerofoil, the lift coefficient can be increased. However, the increase of camber will result in the increase of drag also. Therefore, increased camber is an advantage from lift point of view, but it is a disadvantage from the drag point of view. Therefore, to increase the lift coefficient to achieve low stalling speed by increasing the camber is not an intelligent solution since such a design would result in high drag not only at low speeds but also at high speeds. This problem is solved by incorporating auxiliary devices, which can be used to increase the maximum lift coefficient when required for low-speed operation and can be rendered ineffective at higher speeds. These auxiliary lift devices are broadly classified in the following two categories:

- Those that alter the geometry of the aerofoil.
- Those that control the behaviour of the boundary layer over the aerofoil.

There is of course some overlapping – some devices employ both of these principles. In the former class, the most common device is the trailing edge flap, which can be deflected downwards to give an increase in the effective camber of the wing, and hence to increase lift, whenever that is required, and returned to the neutral position when this lift increment is not required.

The trailing edge flap has another important application. Such flaps are used to vary the lift produced by the various aerofoils of the aircraft (that is, wing, tailplane, and tail fin) in general conditions of flight, so enabling the aircraft to be controlled. When used in this way, they are called flap controls or *control surfaces*. They consist of the following:

(i) *Flap*: It is a control surface used to increase the lift on the wing.
(ii) *Elevator*: It is a control surface used to vary the lift on the horizontal tailplane and so to control the pitching moment.
(iii) *Ailerons*: These are the control surfaces used to vary differentially the lift on the wings and so to control the rolling moment. Ailerons are located at the rear portions of the wing tips.
(iv) *Rudder*: This is a control surface meant for varying the side force on the vertical tail fin and so to control the yawing moment.

Introduction to Aerospace Engineering: Basic Principles of Flight, First Edition. Ethirajan Rathakrishnan.
© 2021 John Wiley & Sons, Inc. Published 2021 by John Wiley & Sons, Inc.
Companion Website: www.wiley.com/go/Rathakrishnan/IntroductiontoAerospaceEngineering

5.2 The Trailing Edge Flap

Trailing edge flap is a small auxiliary aerofoil, located near the rear of the wing, which can be deflected about a given line, where it is hinged. The flap deflection modifies the geometry of the aerofoil, resulting in increased camber, leading to higher lift. A flap used as high-lift device usually deflects only in the downward direction, though the amount of deflection is variable. For a flap designed as a control surface, deflection in both sense is possible, though the range of deflection is usually much less. Another important difference between the *high-lift flap* and *control flap* is that the high-lift device is a cambered profile and the control flap is a symmetrical aerofoil.

5.3 The Plain Flap

A plain flap is illustrated in Figure 5.1. The main effect caused by flap deflection is an increase in the effective camber of the wing. This increase of camber reduces the zero-lift incidence, without affecting the lift-curve slope. Thus, at incidences well below the stall, there is a constant increment in the lift coefficient. This effect may be slightly enhanced because flap deflection additionally produces a slight increase in the effective incidence. Another effect of the flap is that its deflection delays the onset of the stall.

The effect of flap deflection on the lift curve is illustrated in Figure 5.2. It is seen that, when the flap is in neutral position, the lift coefficient, C_L, increases linearly with the angle of attack, α, up to the stalling limit, α_s. For increase of α beyond α_s, C_L decreases steeply, as shown in the figure. The lift coefficient becomes a maximum, $C_{L_{max}}$ at α_s. Beyond α_s, C_L drops suddenly, leading to the stalling of the aircraft.

When the flap is deflected, for a given deflection angle of the flap, the lift coefficient C_L, for all values of α, assumes value higher than that for the neutral flap, as shown in Figure 5.2. Thus, with deflected flap, the aircraft can fly with a considerably reduced flight speed than the flap neutral position without stalling.

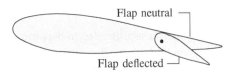

Figure 5.1 An aircraft wing with flap.

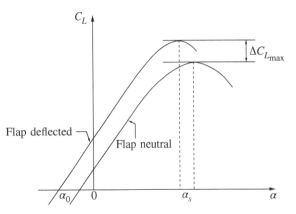

Figure 5.2 Effect of flap deflection on lift curve.

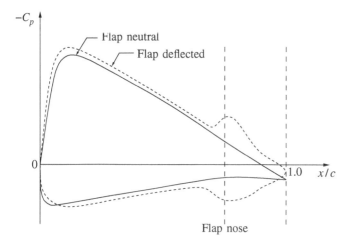

Figure 5.3 Pressure distribution around the wing with flap in neutral and deflected positions.

From Figure 5.2, the following can be inferred:

- The increment in $C_{L_{max}}$ at high incidences is generally less than that at lower incidences, because of the reduced stalling angle. However, the increment in $C_{L_{max}}$ is of primary interest since it is an important parameter, which determines the reduction in stalling speed with flap deflection.
- Further increase in flap deflection generally increases the effects already described; although at high flap deflections the rate of increase of the lift increment begins to fall off for a plain flap, the maximum increment is usually obtained at a flap deflection of about 80°.
- The reduction in stalling angle is *beneficial* because it decreases the incidence during take-off and landing. The aircraft operates in a less nose-up attitude, and the pilot's range of vision is improved.

The effect of flap deflection on the pressure distribution around an aerofoil is shown in Figure 5.3. It is seen that the flap affects the pressure distribution not only at the rear, where the flap is located, but also over the front of the wing profile. The effect of flap deflection in the total increment in lift is clearly seen. It is also seen that much of the lift increment is located over the rear of the wing. This causes the centre of pressure to move downward. There is a nose-down (negative) change in the pitching moment, which has to be corrected by the pilot whenever the flap setting is changed.

Another effect of flap deflection is on the drag of the aerofoil (that is, the wing). The increase in the effective camber causes an increase in the drag coefficient, so that the lift-to-drag ratio is reduced. This is useful for landing because it results in a steeper glide path. This increased drag also results in reduced length of the landing run.

5.4 The Split Flap

The split flap is that which forms a part of the bottom portion at the rear of the wing, as shown in Figure 5.4. Deflection of a split flap alters only the lower surface of the aerofoil. That is, in the case of a split flap, only the lower surface of the rear part of the aerofoil is movable, leading to the upper surface geometry unchanged when the flap is deflected. The split flap deflection also increases the effective camber, giving a reduction in zero-lift incidence. The overall effect is similar to that

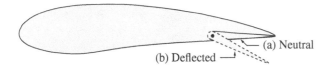

Figure 5.4 Profile with split flap in (a) neutral and (b) deflected positions.

of a plain flap. However, because the upper surface is not so highly cambered, separation effects are less marked so that the performance at high incidence is improved. However, at low incidence the performance is adversely affected because of the large wake behind the deflected flap. This is relatively unimportant since the objective of the flap is to give the wing an improved performance at high incidence.

The effects of split flap deflection are the following:

- The lift-curve slope is slightly increased because the flap performs better at high incidence than at low incidence.
- The zero-lift angle is reduced. However, this decrease is as large as in the case of plain flap.
- The stalling angle is less than with flap neutral but higher than the corresponding value for a plain flap.
- The increment in the maximum lift coefficient is larger than that with a plain flap.
- The effect on pitching moment is similar to that for a plain flap (that is, there will be a negative [−ve] increment) but rather less marked.
- The drag increase caused by the split flap is larger than the plain flap because of the larger wake associated with the split flap.

5.5 The Slotted Flap

In the case of a slotted flap, a gap or slot is opened up between the flap and the main wing when the flap is deflected, as illustrated in Figure 5.5.

The air beneath the wing is at a higher pressure than the air above it. Consequently air blows through the slot, when it is opened, on to the upper surface of the flap. This flow from the bottom to top reenergises the boundary layer and tends to prevent separation. In other words, this flap behaves the same way as a plain flap, but the combination of variable geometry with that amounts to a measure of boundary layer control, resulting in further improvement of performance at all incidences. The increment in the maximum lift coefficient is larger than that provided by a plain or split flap. The drag increment, however, is much less, because of the prevention of separation. This also causes the moment effect to be relatively large.

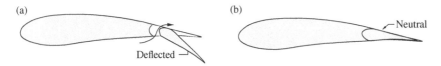

Figure 5.5 A slotted flap in (a) deflected and (b) neutral positions.

Figure 5.6 A Fowler flap in neutral and deflected positions.

5.6   The Fowler Flap

The Fowler flap is similar to the slotted flap, but a further effect is added. In addition to being deflected downwards (camber effect) and opening up of a slot, the flap slides backward, as shown in Figure 5.6. Thus the Fowler flap provides an additional increase in the effective wing area. Therefore, the lift increment caused by the Fowler flap is larger than that provided by other flaps discussed.

There is also a reduction in the effective thickness-to-chord ratio, which tends to make the wing stall a little later. The moment effect is large because the flap, which carries much of the increment, is situated so much to the rear. The drag increment, however, is smaller because of the slot effect and reduced thickness-to-chord ratio.

5.7   Comparison of Different Types of Flaps

We saw that the type of flap has significant effect on the coefficients of lift, drag, and pitching moment. Therefore, to compare the effectiveness of these flaps on the lift coefficient, C_L, its variation with flap deflection angle, α, is given in Figure 5.7 for different types of flaps.

The comparison of the lift curves representing the effects of different types of flap, all at the same moderate flap angle, shows that the deflection of all flaps gives an increased C_L, and this increase of C_L is influenced by the flap type. At a given α, the largest increase in C_L is caused by the Fowler flap.

Figure 5.7 Lift coefficient variation with flap deflection angle.

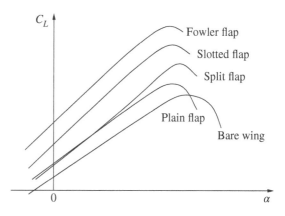

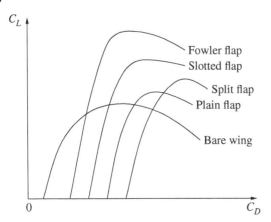

Figure 5.8 Drag polar for different flaps.

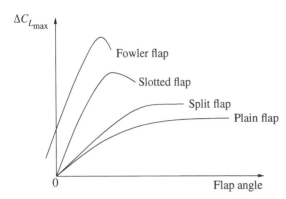

Figure 5.9 Variation of maximum lift coefficient increment with flap deflection angle.

Drag polar for different types of flap is compared in Figure 5.8. The polar curves demonstrate the drag increment for a given lift coefficient. The Fowler flap is seen to produce the largest lift increment for the least drag increment.

Variation of maximum lift coefficient increment with flap angle is shown in Figure 5.9, comparing the performance of different flaps.

The Fowler flap is seen to give a lift increment even at zero flap angle. This is because it slides backwards, giving an increase in effective wing area, even before being deflected downwards. Also, the Fowler flap gives the largest lift increment at any given flap deflection angle. However, the optimum deflection is least in the case of a Fowler flap, and deflection beyond the optimum angle produces a more sudden drop in effectiveness than that associated with other types of flaps.

5.8 Flap Effect on Aerodynamic Centre and Stability

Although flap deflection provides a nose-down change in pitching moment, this is associated simply with a negative (−ve) change in the zero-lift pitching moment coefficient, C_{M_0}. There is generally no significant effect on the derivative dC_M/dC_L and therefore no change in the position of the aerodynamic centre.

Take-off and landing flaps are usually part-span flaps only, situated in the inboard region of the wing, because the corresponding outboard regions are occupied by the ailerons. This reduces the flap effectiveness. It also changes the span-wise load distribution and thus may affect the induced

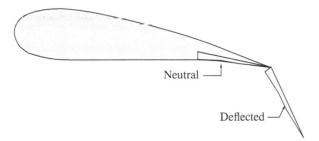

Figure 5.10 A zap flap in neutral and deflected positions.

drag. It has the further effect of moving the centre of lift inwards on each wing, and this may affect the lateral stability of the aircraft.

There are, of course, other types of flap in addition to those discussed above. In general, they consist of the extensions and adaptions of just the same principle:

- *Multiple slotted flaps*: These use the same concept as the single slotted flap but have increased effectiveness.
- *The zap flap*: This is a combination of the split flap concept and the Fowler flap concept. A typical zap flap is illustrated in Figure 5.10.

A zap flap combines a higher-lift increment than that of a split flap with a higher-drag increment than that of a slotted or Fowler flap.

5.9 The Leading Edge Slat

The leading edge slat is an auxiliary aerofoil or slat, mounted in a fixed position ahead of the leading edge of the main aerofoil (that is wing), with a carefully designed gap, or slot, between them, as shown in Figure 5.11. The leading edge slat is essentially a device meant for boundary layer control. It operates by allowing the passage, through the slot, of air from the higher-pressure region below the wing to the lower pressure region above it. Thus, energy is added to the boundary layer on the upper surface, and any tendency to separation of the flow is much reduced.

As the incidence increases, the tendency for the flow to separate from the upper surface of the wing would increase, and thus separation would begin from the leading edge and spread in the downward direction. Positioning a slat at the leading edge, as shown in Figure 5.11, would prevent the flow separation, leading to the continuous increase of lift coefficient with incidence. Because of this, the stalling will be delayed. The linear part of the lift curve will be extended. The stalling angle might be typically increased from, say, 15° to 25°, and the increase in the maximum lift coefficient to the order of 60% is possible. The variation of lift coefficient, C_L, with incidence, α, for a plain wing and a wing with leading edge slat, shown in Figure 5.12 illustrates this behaviour.

Figure 5.11 Sectional view of a wing with leading edge slat.

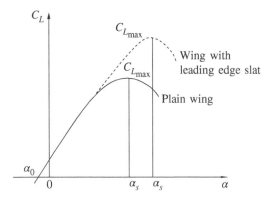

Figure 5.12 Illustration of leading edge slat on $C_{L_{max}}$.

The drag on the wing at low speeds, that is, at high angle of incidence, is greatly reduced by the separation prevention caused by the leading edge slat. However, the effect of the slat on pitching moment is very small.

The increase of $C_{L_{max}}$ obtained with a leading edge slat is comparable with that obtained with a trailing edge flap. However, there are number of disadvantages associated with the use of a fixed slat at the leading edge of the wing. The disadvantages are the following:

- As we saw, the drag at low speeds is reduced, whereas for landing, in particular, high drag and low lift-to-drag ratio is desirable.
- At low incidence, that is, at high speed, the slat ahead of the leading edge tends to spoil the flow over the rest of the wing, and there is an appreciable increase in C_D. Thus, the leading edge slat causes increase in drag at the wrong end of the speed range.
- Maximum lift coefficient is obtained only at very high incidence, so that the attitude of the aircraft at take-off and landing is very much nose-up, and the pilot's range of vision is impaired.

The introduction of a movable slat can overcome the disadvantage of high drag at low incidence, since the slat could be withdrawn and the slot closed at high speeds.

5.10   The Leading Edge Flap

The leading edge flap is another device capable of producing lift increment by preventing flow separation from the leading edge of the wing. The leading edge flap is also referred to as *droop-snoot*.

High-speed aerofoil sections usually have fairly sharp leading edge and very little camber. Such wings tend to stall early because, even at moderate incidence, the flow around the leading edge has to negotiate a sharp bend. This gives rise to a region of separated flow, known as a separation bubble, behind the leading edge, as shown in Figure 5.13.

Not only the separation bubble reduces the lift of the aerofoil, but also, as the incidence increases, the bubble bursts, and the separation spreads downstream, leading to an early and particularly sudden and violent stall. This kind of separation can be prevented by using a certain amount of camber at the leading edge, as shown in Figure 5.14. The flow can negotiate the leading edge smoothly, and the stall is considerably delayed.

However, the leading edge camber tends to increase the drag at low incidence. The drag increase can be avoided by a nose that can be deflected downstream to give the necessary improvement in high incidence performance and retracted so as to give the usual nose shape at low incidence. Such a device is the leading edge flap, illustrated in Figure 5.15.

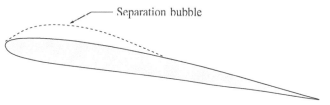

Figure 5.13 A separation bubble just behind the leading edge of an aerofoil.

Figure 5.14 An aerofoil with leading edge camber.

Figure 5.15 An aerofoil with leading edge flap.

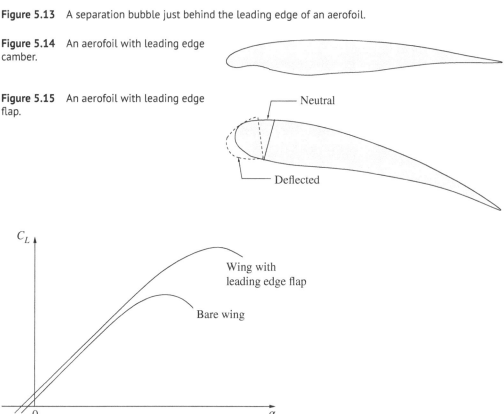

Figure 5.16 Variation of C_L with incidence for wings with and without leading edge flap.

The leading edge flap could be used on low-speed sections, but it is mainly applied to high-speed aerofoils for the reasons given above. Its effect on the lift curve is similar to that of a leading edge slot, but it results in reduced camber effect as well. The lift coefficient variation with angle of incidence for wings with and without leading edge flap is compared in Figure 5.16. The stall delay caused by leading edge flap is not so large as that of leading edge slot.

5.11 Boundary Layer Control

Boundary layer control is a method of increasing the lift generated by a wing, without changing its geometry. Thus boundary layer control is an artificial high-lift device. The aim is to prevent or delay separation by blowing air into the boundary layer to energise it or sucking away the boundary layer.

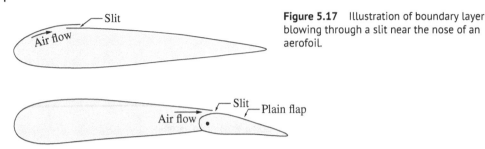

Figure 5.17 Illustration of boundary layer blowing through a slit near the nose of an aerofoil.

Figure 5.18 A blown flap ahead of a flap.

5.11.1 Boundary Layer Blowing

The principle of boundary layer blowing is similar to that of the leading edge slot. High-speed air is blown into the boundary layer through a narrow slit, as illustrated in Figure 5.17, in the upper surface of the wing, where it reenergises the boundary layer and prevents separation. Since the velocity of the air fed in this way is so much higher than the speed of air passing through the leading edge slot, or a slotted flap, blowing will generally prove to be much more effective. The stall can be delayed almost indefinitely by this means. In addition, the jet of air has the effect of increasing the circulation round the wing, thus giving a direct lift increment at all incidences.

The slit to blow the boundary layer may be near the nose of the aerofoil, as illustrated in Figure 5.17, so that the blowing affects the whole of the upper surface. Alternatively, the slot may be located just upstream of the nose of a plain flap, as shown in Figure 5.18. In this position, the flow upstream of the slot will be affected to some extent by induction, but the main objective is to prevent separation of the flow over the upper surface of the flap. This device is known as the blown flap, as illustrated in Figure 5.18.

There is some advantage in this device compared with a slot placed further forward. In the latter case, the effect of blowing may be reduced by the time the flow reaches the rear of the aerofoil, where separation is most likely.

5.12 Boundary Layer Suction

The principle of suction of boundary layer is the removal of slowly moving air in the boundary layer, so that there is no layer to separate. A series of holes, flushed with the surface, are made in the surface of the aerofoil upstream of the separation point, as shown in Figure 5.19, and the air in the boundary layer is sucked into the wing through these holes. However, from this point onwards, the boundary layer will re-form and thicken, and separation may still occur at some point downstream. To prevent this, a series of suction holes must be made at various chord-wise positions, as shown in Figure 5.19. The logical extension of this idea is the use of a porous wing surface, with suction applied everywhere on the surface.

In addition to preventing separation, suction may also be used to prevent transition and hence to keep drag low. Such a device would appear to be of particular interest in conjunction with the use of low-drag wing sections. The principle behind the design of a low drag section is the maintenance of laminar flow. The disadvantage associated with such a design is that separation occurs easily when the incidence is increased even by a small amount above the design value.

Figure 5.19 An aerofoil with suction ports over its upper surface.

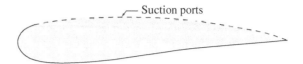

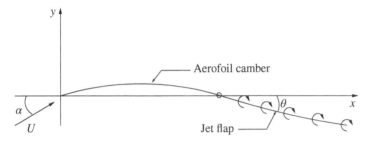

Figure 5.20 A jet flap.

5.13 The Jet Flap

The jet flap consists of a very high-speed jet of air blown out through a narrow slit in the trailing edge of the wing, as illustrated in Figure 5.20.

The jet, deflected slightly downwards, divides the upper surface flow from the lower surface flow and produces an effect on the flow over the wing similar to that produced by a very large physical trailing edge flap. There is an additional increment in lift due to the downward component of the momentum of the jet, leading to the very high-lift coefficient.

5.14 Summary

The lift coefficient of an aerofoil is strongly influenced by its camber. The higher the camber, the high the lift coefficient.

Auxiliary lift devices are broadly classified in the following two classes: those that alter the geometry of the aerofoil and those that control the behaviour of the boundary layer over the aerofoil. The control surfaces consist of the following:

Flap   is a control surface used to increase the lift on the wing.
Elevator is a control surface used to vary the lift on the horizontal tailplane and so to control the pitching moment.
Ailerons are control surfaces used to vary differentially the lift on the wings and so to control the rolling moment. These are located at the rear portions of the wing tips.
Rudder   is a control surface meant for varying the side force on the vertical tail fin and so to control the yawing moment.

Trailing edge flap is a small auxiliary aerofoil, located near the rear of the wing, and which can be deflected about a given line, where it is hinged.

An important difference between the *high-lift flap* and *control flap* is that the high-lift device is a cambered profile and the control flap is a symmetrical aerofoil.

The main effect caused by flap deflection is an increase in the effective camber of the wing. When the flap is deflected, for a given deflection angle of the flap, the lift coefficient C_L, for all values of α, assumes value higher than that for the neutral flap.

The flap affects the pressure distribution not only at the rear, where the flap is located, but also over the front of the wing profile. Another effect of flap deflection is on the drag of the wing. This increased drag also results in reduced length of the landing run.

The split flap is that which forms a part of the bottom portion at the rear of the wing. Deflection of a split flap alters only the lower surface of the aerofoil.

The effects of split flap deflection are the following. The lift-curve slope is slightly increased, because the flap performs better at high incidence than at low incidence. The zero-lift angle is reduced. However, this decrease is as large as in the case of plain flap. The stalling angle is less than with flap neutral but higher than the corresponding value for a plain flap. The increment in the maximum lift coefficient is larger than that with a plain flap. The effect on pitching moment is similar to that for a plain flap (that is, there will be a negative [−ve] increment) but rather less marked. The drag increase caused by the spit flap is larger than the plain flap because of the larger wake associated with the split flap.

In the case of a slotted flap, a gap or slot is opened up between the flap and the main wing when the flap is deflected.

The Fowler flap is similar to the slotted flap, but a further effect is added. In addition to being deflected downwards (camber effect) and opening up of a slot, the flap slides backward.

Take-off and landing flaps are usually part-span flaps only, situated in the inboard region of the wing, because the corresponding outboard regions are occupied by the ailerons.

Multiple slotted flaps use the same concept as the single slotted flap but have increased effectiveness.

The zap flap is a combination of the split flap concept and the Fowler flap concept.

The leading edge slat is an auxiliary aerofoil or slat, mounted in a fixed position ahead of the leading edge of the main aerofoil (that is wing), with a carefully designed gap, or slot, between them.

The disadvantages associated with the use of a fixed slat at the leading edge of the wing are the following. The drag at low speeds is reduced, whereas for landing, in particular, high drag and low lift-to-drag ratio is desirable. At high speed, the slat ahead of the leading edge tends to spoil the flow over the rest of the wing, and there is an appreciable increase in C_D. Thus, the leading edge slat causes increase in drag at the wrong end of the speed range. Maximum lift coefficient is obtained only at very high incidence, so that the attitude of the aircraft at take-off and landing is very much nose-up, and the pilot's range of vision is impaired.

Introduction of a movable slat can overcome the disadvantage of high drag at low incidence, since the slat could be withdrawn and the slot closed at high speeds.

Leading edge flap is another device capable of producing lift increment by preventing flow separation from the leading edge of the wing.

High-speed aerofoil sections usually have fairly sharp leading edge and very little camber. Such wings tend to stall early because, even at moderate incidence, the flow around the leading edge has to negotiate a sharp bend. This gives rise to a region of separated flow, known as a separation bubble, behind the leading edge.

The leading edge flap could be used on low-speed sections, but it is mainly applied to high speed aerofoils for the reasons given above. Its effect on the lift curve is similar to that of a leading edge slot.

Boundary layer control is a method of increasing the lift generated by a wing, without changing its geometry.

The principle of boundary layer blowing is similar to that of the leading edge slot.

The principle of suction of boundary layer is the removal of slowly moving air in the boundary layer, so that there is no layer to separate. In addition to preventing separation, suction may also be used to prevent transition and hence to keep drag low.

The jet flap consists of a very high-speed jet of air blown out through a narrow slit in the trailing edge of the wing.

6

Thrust

6.1 Introduction

Thrust is the force that opposes drag and enables the airplane going forward. In a steady level flight, the thrust must be equal to the drag. To accelerate the aircraft the thrust must be greater than the drag. In climbing flight also, the thrust must be greater than the drag. The performance of the aircraft largely depends on the amount of thrust provided by its engines.

An aircraft engine takes in air through its intake, fuel is added to the air, the fuel–air mixture is burned in the combustion chamber, and the combustion products at high pressure and high temperature are accelerated through the nozzle of the engine, discharging to the atmosphere. Thrust generated by an engine relies on the reaction of the gas ejected through the engine. The physical process by which this reaction is produced and transmitted to the aircraft depends on the type of propulsion system used.

For all types of propulsion system, the energy is supplied in the form of a fuel, which is fed into the combustion chamber, where it is burned. The thermal energy produced by burning the fuel and air is converted into the mechanical work done in propelling the aircraft against the drag.

The piston engine and propeller are the oldest among the propulsion systems. The piston engine is still in use for light aircraft and involves a large number of mechanical parts. In the ramjet and rocket engines, there are only a few significant mechanical parts, apart from fuel pumps, and the entire engine relies completely on aerodynamic and thermodynamic principles. Between the two extremes there are turbojet and turboprop engines that involve a combination of mechanical and aerothermodynamic principles.

6.2 Thrust Generation

Thrust is the forward force that pushes the engine and, therefore, the airplane forward. Sir Isaac Newton discovered that for 'every action there is an equal and opposite reaction'. An engine uses this principle. The engine takes in a large volume of air. The air is decelerated and compressed and slowed down. The air is forced through many spinning blades. By mixing this air with jet fuel and burning the fuel–air mixture, the temperature of the burnt gas can be increased to as high as 3000 K. The power of this gas is used to turn the turbine. Finally, when the air leaves, it is pushed backward out of the engine. This causes the airplane to move forward.

Jet engines move the airplane forward with a great force that is produced by a tremendous thrust and causes the airplane to fly very fast.

Introduction to Aerospace Engineering: Basic Principles of Flight, First Edition. Ethirajan Rathakrishnan.
© 2021 John Wiley & Sons, Inc. Published 2021 by John Wiley & Sons, Inc.
Companion Website: www.wiley.com/go/Rathakrishnan/IntroductiontoAerospaceEngineering

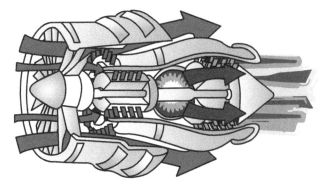

Figure 6.1 Illustration of flow through gas turbine engine.

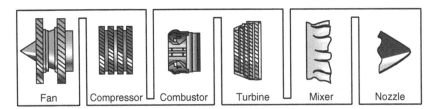

Figure 6.2 Schematic diagram illustrating the components of a jet engine.

All jet engines, which are also called *gas turbines*, work on the same principle. The engine sucks air with a fan. A compressor raises the pressure of the air, and it is made with many blades attached to a shaft. The blades spin at high speed and compress or squeeze the air. The compressed air is then sprayed with fuel, and an electric spark lights the mixture. The burning gases expand and discharged out through the nozzle, at the back of the engine. As the jet of gas shoots backward, thrust is produced. This thrust moves the aircraft forward. As the hot air is going through the nozzle, it passes through another group of blades called the turbine. The turbine is attached to the same shaft as the compressor. Spinning the turbine causes the compressor to spin.

The air goes through the core of the engine as well as around it. This causes some of the air to be very hot and some to be cooler. The cooler air then mixes with the hot air at the engine exit area, as illustrated in Figure 6.1.

The major parts of a jet engine are fan, compressor, combustion chamber, turbine, mixer, and nozzle. Artistic views of the engine parts are shown in Figure 6.2:

- *Fan*: The fan is the first component in a turbofan engine. The large spinning fan sucks in large quantity of air. Most blades of the fan are made of titanium. It then speeds up this air and splits it into two parts. One part continues to flow through the 'core' or centre of the engine, where it is acted upon by the other engine components. The second part 'bypasses' the core of the engine. It goes through a duct that surrounds the core to the back of the engine, where it produces much of the force that propels the airplane forward. This cooler air helps to quieten the engine in addition to adding thrust to the engine.
- *Compressor*: The compressor is the first component in the engine core. It is made up of fans with many blades and attached to a shaft and squeezes the air that enters it into progressively smaller areas, resulting in an increase in the air pressure. This results in an increase in the energy potential of the air. The squashed air is forced into the combustion chamber.

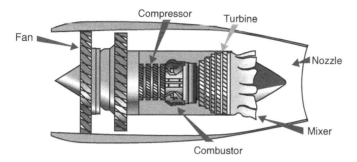

Figure 6.3 A view of complete engine.

- *Combustor*: In the combustor, the air is mixed with fuel and then ignited. There are as many as 20 nozzles to spray fuel into the airstream. The mixture of air and fuel catches fire. This provides a high-temperature, high-energy airflow. The fuel burns with the oxygen in the compressed air, producing hot expanding gases. The inside of the combustor is often made of ceramic materials to provide a heat-resistant chamber. The heat of the combustion product can reach about 2700 °C.
- *Turbine*: The high-energy airflow coming out of the combustor goes into the turbine, causing the turbine blades to rotate. The turbines are linked by a shaft to turn the blades in the compressor and to spin the intake fan at the front. This rotation takes some energy from the high-energy flow that is used to drive the fan and the compressor. The gases produced in the combustion chamber move through the turbine and spin its blades. The turbines of the jet engine spin around thousands of times per second. (For example, large jet engines operate around 10 000–25 000 rpm, while microturbines spin as fast as 500 000 rpm.) They are fixed on shafts that have several sets of ball bearing in between them.
- *Nozzle*: The nozzle is the exhaust duct of the engine. This is the engine part that actually produces the thrust for the airplane. The energy depleted airflow that passed the turbine, in addition to the colder air that bypassed the engine core, produces a force when exiting the nozzle that acts to propel the engine, and therefore the airplane moves forward. The combination of the hot air and cold air are expelled and produce an exhaust, which causes a forward thrust. The nozzle may be preceded by a mixer, which combines the high-temperature air coming from the engine core with the lower temperature air that was bypassed in the fan. The mixer helps to make the engine quieter.

A view of complete gas turbine engine is shown in Figure 6.3.

6.2.1   History of Jet Engine Development

Sir Isaac Newton in the eighteenth century was the first to theorise that a rearward-channelled explosion could propel a machine forward at a great rate of speed. This theory was based on his third law of motion. As the hot air blasts backwards through the nozzle, the plane moves forward. Henri Giffard built an airship that was powered by the first aircraft engine, a three-horse power steam engine. It was very heavy, too heavy to fly. In 1874, Félix du Temple built a monoplane that flew just a short hop down a hill with the help of a coal fired steam engine. In the late 1800s, Otto Daimler invented the first gasoline engine. In 1894, American Hiram Maxim tried to power his triple biplane with two coal fired steam engines. It only flew for a few seconds. The early steam engines were powered by heated coal and were generally too heavy for flight. American Samuel Langley made a model airplanes that were powered by steam engines. In 1896, he was successful in flying an unmanned airplane with a steam-powered engine, called the Aerodrome. It flew about 1 mi before it ran out of steam. He then tried to build a full-sized plane, the Aerodrome A, with a gas-powered engine. In 1903, it crashed immediately after being launched from a house boat.

In 1903, the Wright Brothers flew The Flyer with a 12-horse power gas-powered engine. From 1903, the year of the Wright Brothers first flight, to the late 1930s, the gas-powered reciprocating internal combustion engine with a propeller was the sole means used to propel aircraft. It was Frank Whittle, a British pilot, who designed and patented the first turbojet engine in 1930. The Whittle engine first flew successfully in May 1941. This engine featured a multistage compressor and a combustion chamber, a single-stage turbine, and a nozzle. At the same time that Whittle was working in England, Hans von Ohain was working on a similar design in Germany. The first airplane to successfully use a gas turbine engine was the German Heinkel He 178 in August 1939. It was the world's first turbojet-powered flight. General Electric built the first American jet engine for the US Army Air Force jet plane. It was the XP-59A experimental aircraft that first flew in October 1942.

6.2.2 Types of Jet Engines

Jet engines can be broadly classified as turbojets, turboprops, turbofans, turboshafts, and ramjets.

6.2.2.1 Turbojets

The basic idea of the *turbojet engine* is simple. Air taken in from an opening in the front of the engine is compressed to 3–12 times its original pressure in compressor. Fuel is added to the air and burned in a combustion chamber to raise the temperature of the fluid mixture to about 1100–1300 °F. The resulting hot air is passed through a turbine, which drives the compressor. If the turbine and compressor are efficient, the pressure at the turbine discharge will be nearly twice the atmospheric pressure, and this excess pressure is sent to the nozzle to produce a high-velocity stream of gas that produces a thrust. Substantial increase in thrust can be obtained by employing an afterburner. It is a second combustion chamber positioned after the turbine and before the nozzle. The *afterburner* increases the temperature of the gas ahead of the nozzle. The result of this increase in temperature is an increase of about 40% in thrust at take-off and a much larger percentage at high speeds once the plane is in the air.

The turbojet engine is a reaction engine. In a reaction engine, expanding gases push hard against the front of the engine. The turbojet sucks in air and compresses or squeezes it. The gases flow through the turbine and make it spin. These gases bounce back and shoot out of the rear of the exhaust, pushing the plane forward. A pictorial view of turbojet engine is shown in Figure 6.4.

6.2.2.2 Turboprops

A *turboprop engine* is a jet engine attached to a propeller. The turbine at the back is turned by the hot gases, and this turns a shaft that drives the propeller. Some small airliners and transport aircraft are powered by turboprops.

Like the turbojet, the turboprop engine consists of a compressor, combustion chamber, and turbine. The air and gas pressure is used to run the turbine, which then creates power to drive the compressor. Compared with a turbojet engine, the turboprop has better propulsion efficiency at flight speeds below about 800 km/h. Modern turboprop engines are equipped with propellers that have a smaller diameter but a larger number of blades for efficient operation at much higher flight speeds. To accommodate the higher flight speeds, the blades are scimitar shape (shaped like a scimitar sword, with increasing sweep along the leading edge) with swept-back leading edges at the blade

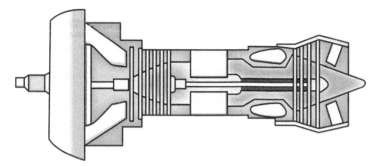

Figure 6.4   A view of turbojet engine.

Figure 6.5   A view of turboprop engine.

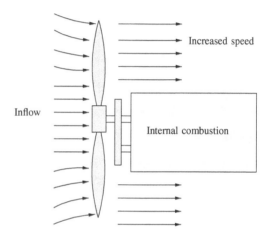

tips. Engines featuring such propellers are called *prop-fans*. An artistic view of turboprop engine is shown in Figure 6.5.

6.2.2.3   Turbofans

A turbofan engine has a large fan at the front, which sucks in air. Most of the air flows around the outside of the engine, making it quieter and giving more thrust at low speeds. Most of today's air-liners are powered by turbofans. In a turbojet, all the air entering the intake passes through the gas generator, which is composed of the compressor, combustion chamber, and turbine. In a turbofan engine, only a portion of the incoming air goes into the combustion chamber. The remainder passes through a fan or low-pressure compressor and is ejected directly as a 'cold' jet or mixed with the gas-generator exhaust to produce a 'hot' jet. The objective of this sort of bypass system is to increase thrust without increasing fuel consumption. This is achieved by increasing the total air mass flow and reducing the velocity within the same total energy supply. An artistic view of turbofan engine is shown in Figure 6.6.

6.2.2.4   Turboshafts

This is another form of gas turbine engine that operates much like a turboprop system. It does not drive a propeller. Instead, it provides power for a helicopter rotor. The turboshaft engine is designed so that the speed of the helicopter rotor is independent of the rotating speed of the gas generator. This permits the rotor speed to be kept constant even when the speed of the generator is

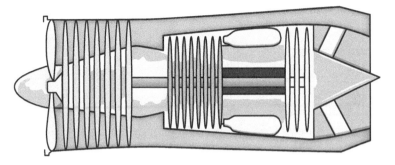

Figure 6.6 A view of turbofan engine.

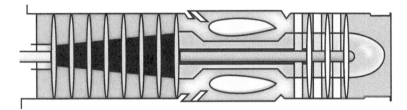

Figure 6.7 A view of turboshaft engine.

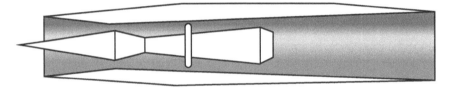

Figure 6.8 A view of ramjet engine.

varied to modulate the amount of power produced. An artistic view of turboshaft engine is shown in Figure 6.7.

6.2.2.5 Ramjets
The ramjet is the simplest jet engine and has no moving parts. The speed of the jet 'rams' or forces air into the engine. It is essentially a turbojet in which rotating machinery has been omitted. Its application is restricted by the fact that its compression ratio depends wholly on forward speed. The ramjet develops no static thrust and very little thrust in general below the speed of sound. As a consequence, a ramjet vehicle requires some form of assisted take-off, such as another aircraft. It has been used primarily in guided-missile systems. Space vehicles use this type of jet engine. An artistic view of ramjet engine is shown in Figure 6.8.

6.3   Turbojet

The turbojet, usually referred to as the *jet engine*, is the commonly used aircraft engine at present. In principle, the turbojet is a very simple form of propulsion unit based on the *gas turbine engine*, illustrated in Figure 6.9.

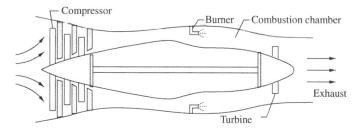

Figure 6.9 Schematic diagram illustrating the components of a jet engine.

The pressure and temperature of the air entering the engine are controlled by the action of the compressor. The air at low speed entering the combustion chamber is mixed with fuel, and the fuel–air mixture is burned. The fuel used is usually kerosene, also termed *aviation gasoline*. Heat added by the combustion increases the pressure and temperature of the combustion product. The combustion product at high pressure leaves the engine as a high-speed jet. As the speed of the jet increases, the momentum increases, and the reaction to this momentum change is a thrust force pushing the engine forward. The intake of the engine faces forward; thus in flight the air is effectively 'rammed' in. This *ram effect* helps to compress the air. In addition, as the forward speed increases, less and less work has to be done by the compressor, leaving more of energy to be used to generate thrust, leading to the increased efficiency of the propulsion system.

The turbojet has no reciprocating parts, such as the piston-cylinder in the piston engine. Because of the absence of reciprocating parts, the wear and vibrations associated with turbojet are insignificant. Another important advantage is that the turbojet produces significantly larger thrust for a given weight at high speed, compared with a piston engine. Furthermore, it will work efficiently close to and beyond speed of sound, where propellers cannot be used.

6.4 Turboprop and Turboshaft Engines

The construction of turboprop engine is almost the same as the turbojet engine, but more of the available energy in the exhaust is used to drive the turbine. The extra power produced by the turbine is used to drive the propeller. Some thrust is produced by the exhaust jet, but this is only relatively a small portion of the total thrust produced. The advantage of the turboprop engine over the turbojet engine is that it is much more efficient. As it will be seen later, the turboprop engine is more efficient to produce thrust by giving a small increase in momentum to a large amount of air (as in the case of a propeller engine) than to give a large increase of momentum to a small amount of air (as in a turbojet engine). However, propellers produce serious problems at high speed. Also, they are noisy and require high maintenance and the addition of a heavy gearbox to reduce the turbine speed, which can be up to 100 000 rpm or more, down to a few 1000 rpm required for a propeller.

Instead of driving a propeller, a gas turbine may be used to drive the rotor blades of a helicopter, and in this application it is normally known as a *turboshaft* engine. Most military helicopter engines are turboshaft engines. These engines are also used for the production of auxiliary electrical power.

6.5 Ramjet and Scramjet

The ramjet is used for propelling missiles and early stages of spacecraft flight, although not used for aircraft propulsion. A ramjet is a simple air-breathing engine. It consists of a diffuser, a combustion

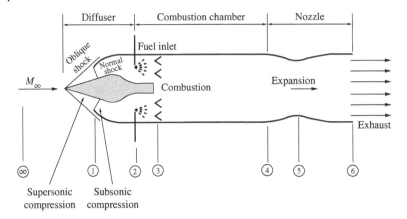

Figure 6.10 Schematic of a ramjet engine. (∞) Freestream; (1) oblique shock; (2) fuel spray; (3) flame-holder; (4) nozzle entry; (5) nozzle throat; (6) nozzle exit.

chamber, and an exhaust nozzle, as shown in Figure 6.10. Air entering the diffuser is compressed to attain the low-speed level required for combustion. Fuel is mixed with air in the combustion chamber and burned. The hot gas mixture, after combustion, is expanded through the nozzle that converts the high-pressure energy of the combustion gases to kinetic energy, resulting in the generation of thrust. It is important to note that, in the diffuser, the incoming air is decelerated from the high flight speed M_∞, which is usually supersonic, to a relatively low velocity through the compression caused by the shock system at the intake and the ramming action inside the duct from just downstream of the inlet to the combustion chamber entrance. Therefore, although ramjets can operate at subsonic flight speeds, the increasing pressure rise accompanying higher flight speeds renders ramjets more suitable for supersonic flights.

The ramjet engine shown in Figure 6.10 is typical of supersonic ramjets that employ supersonic diffusion through a system of shocks. Since the supersonic flow stream entering the intake has to be decelerated to about Mach 0.2–0.3 before entry to the combustion chamber, the pressure rise can be substantial. For example, for isentropic deceleration from Mach 4 to 0.3, the static pressure ratio between the ambient and the combustion chamber should be around 145. However, in reality, only a fraction of the isentropic pressure ratio is actually achieved. This is because, at high Mach numbers, the total pressure losses associated with shocks are substantial.

After compression the air passes over the fuel injectors that spray fuel in the form of fine droplets. The fuel and air mix rapidly, and the mixture is then burned in the combustion chamber. A flame holder is placed in the combustion chamber to stabilise the flame. The combustion raises the temperature of the fuel–air mixture to about 3000 K before the combustion products expand to a high velocity in the nozzle. The reaction caused by the momentum of the combustion products leaving the nozzle is the thrust generated by the engine. The thrust (T) generated can be expressed as

$$T = \dot{m}\, u_e + \left(p_e - p_\infty\right) A_e$$

where $\dot{m}$ is the mass flow rate through the nozzle, u_e is the flow velocity at the nozzle exit, p_e is the pressure at the nozzle exit, p_∞ is the ambient pressure to which the nozzle discharges, and A_e is the nozzle exit area. A part of this thrust is used to overcome the shear force acting over the internal surface of the engine.

As we know, gas turbine engines cannot be operated at high temperatures of the order of 3000 K. This is because the turbine blades are subjected to high centrifugal stresses and cannot be cooled easily. However, the walls of the combustion chamber and the nozzle of a ramjet engine can be cooled by a fuel-injection pattern that leaves a shielding layer of relatively cool air next to the wall surface to maintain the wall temperature around 1000 K, which can be tolerated by currently available materials. This relatively higher operating temperature of the ramjet allows operation at high flight Mach numbers. However, the combustion chamber inlet temperature also increases with flight Mach number. This sets a limiting Mach number for ramjet operation, since beyond the limiting Mach number the temperature will approach the limit set by wall material and cooling methods. For example, at a flight Mach number of 8 in an atmosphere at 225 K, the stagnation temperature is about 2500 K.

At temperatures of more than 2500 K, dissociation of combustion products becomes significant. At such temperatures further addition of fuel will result in dissociation enhancement rather than temperature rise.

A major disadvantage of the ramjet is the pressure rise in the engine, and this is solely governed by the flight speed and the diffuser performance. Owing to this the ramjet cannot develop static thrust and hence cannot accelerate a vehicle from zero velocity. Also, designing a diffuser of high efficiency is difficult due to detrimental boundary layer behaviour in raising the pressure gradients especially in the presence of shocks, which cannot be avoided during supersonic operation, since a supersonic flow is essentially wave dominated. However, the complicated thermodynamic process associated with the ramjet can be greatly simplified if certain assumptions are made. A ramjet designed with such simplified thermodynamic analysis is termed an *ideal ramjet*.

6.6 The Ideal Ramjet

It is essentially a hypothetical engine with simplified cycles involving isentropic processes. The complicated thermodynamic processes involved in the ramjet flow and combustion can be greatly simplified by making the following assumptions:

- The compression and expansion processes in the engine are reversible and adiabatic.
- The combustion process takes place at constant pressure.

Even though these assumptions are not realistic, they simplify the analysis of the processes in the ideal ramjet engine. Furthermore, the ideal ramjet is a useful concept and serves as a standard for comparing the performance of actual ramjet engines, that is, the ideal ramjet is equivalent to the Carnot engine model used in thermodynamics. In the diffusers of the actual engines, there are irreversibilities due to the shock waves, fuel–air mixing, and wall friction. Also, unless the combustion takes place at very low velocities, both static and total pressures will drop due to friction and heat transfer.

In the ramjet engine shown in Figure 6.10, the air is isentropically compressed from the free stream state ∞ to stagnation state 02 at station 2, as shown in the *T–s* diagram in Figure 6.11. It is essential to realise that in the ideal process assumed, the compression through the oblique shock and normal shock from stations 1 to 2 is regarded as isentropic. This is reasonable only when the oblique shock at the nose of the intake spike is very weak, causing flow turning of less than 5°, and the Mach number upstream of the normal shock is just about unity. Even after meeting these stringent requirements, the compression process is only approximately isentropic. In spite of these assumptions, of weak oblique and normal shocks, the actual process is nonisentropic. However, the entropy increase associated with the compression process would be small enough for us to approximate it as isentropic. The combustion process is represented by a constant-pressure heat and mass addition process from state 02 to state 04, where the temperature is a maximum at T_{04}.

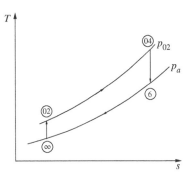

Figure 6.11 *T–s* diagram of an ideal ramjet process.

The combustion products expand isentropically through the nozzle to the ambient pressure p_∞ to which the flow is discharged. That is, in the ideal process the nozzle is assumed to be correctly expanded with the exit pressure p_e equal to the backpressure p_∞.

6.7 Rocket Propulsion

Rocket propulsion is straightforward in principle. As illustrated in Figure 6.12, in the simplest rocket motors, fuel is burned in a combustion chamber to create heat and a high-pressure gas. The hot gas then flows out through the specially shaped nozzle at high speed. The main difference between the rocket and other forms of propulsion is that *air is not used as the oxidant in the burning process, and the gases that are emitted from the outlet are all derived from the fuel*. In other words, in a rocket motor, the fuel and oxidiser are stored in the motor itself. Thus, a rocket is essentially a non-air-breathing engine unlike a turbine engine that takes in air from the atmosphere as the oxidiser. The rocket is mainly used for propelling missiles and spacecraft.

Rocket motors come in two basic types, namely, solid fuel type and liquid propellant type. In solid fuel type, the fuel and oxidiser are combined in a stable solid form, as in a simple firework. In a liquid propellant type, two chemicals, one usually the fuel and the other the oxidiser, are mixed together and burned in a combustion chamber.

Rocket motors have a very high thrust-to-weight ratio and are essentially very simple. However they use fuel and oxidiser at a very high rate and so have relatively short duration. The need to carry oxidiser as well as fuel means that the total weight of expendable chemicals carried is much higher than in engines that use air as the oxidant. Recently, engines have been developed that use air as the oxidant at low altitudes and pure rocket propulsion at high altitude or in space.

Apart from the simple chemically fuelled rockets, several more advanced types of rockets have been developed. These include the rockets in which ionised particles or plasma is accelerated to very

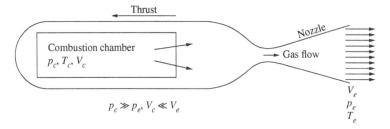

Figure 6.12 A simple rocket motor.

Figure 6.13 A typical propeller engine.

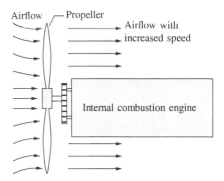

high speeds by electrostatic or electromagnetic forces. Some of these have been used on spacecraft, normally as low-thrust control jets.

6.8 Propeller Engines

Propeller engines are essentially internal combustion engines. The propeller in the engine driven by the internal combustion engine takes in a large air mass and pushes it with a marginally increased speed. Because of the large amount of air mass, the thrust generated becomes significant, even though the velocity increase induced to the airflow is small. In a propeller engine there is a clear dividing line between the propeller and the engine, as illustrated in Figure 6.13.

In a propeller engine, the engine that drives the propeller may itself be a gas turbine, and in this case the designer has the choice to allot the proportion of the power delivered by the propeller and jet as per the design considering as in the *turboprop system.*

6.9 Thrust and Momentum

All propulsive systems deliver thrust as a result of giving momentum to air or other gases. Thus the *amount of thrust generated* will be equal to the rate at which *momentum is imparted to the air.*

If $\dot{m}$ is the mass flow rate of air and V is the velocity imparted to the air by the propulsion device, the thrust generated is

$$T = \dot{m}\ V$$

From this expression for thrust, it is evident that the thrust generated can be controlled by the adjustments of either mass flow rate or velocity. The propeller engine throws back a large mass of air with comparatively low velocity increase; on the other hand, the jet engine throws back a relatively smaller amount of mass with a very high-velocity increment.

At this stage we may ask, if the propeller engine generates thrust by simply handling more mass and imparting a small velocity increment than a jet engine, why then is the propeller less popular than jet engine? The answer to this question is the following:

- Since the propeller has a rotational as well as a forward speed, it follows that the blade tips will start to move through the air faster than the speed of sound long before the rest of the aircraft. The occurrence of supersonic flow at the blade tips causes all sorts of problems. Because of these problems, the propeller engines are less popular than the jet engines that provide the only practical alternative for high-speed flight.

6.10 Bypass and Turbofan Engines

A jet engine generates thrust by giving a large increase in speed to a relatively smaller mass than a piston or propeller engine. Therefore, to make a jet engine more efficient, the design should be such that a larger mass of air is somehow given a smaller increase in speed. The method used is to increase the size of the compressor fan to allow a portion of the air to pass around the outside of the engine. There are also a number of secondary advantages, the most significant being a reduction in noise. Another function of bypass air is to cool the engine and to make use of some of the otherwise wasted heat to increase the thrust.

By increasing the amount of bypass air, the so-called fan jet, illustrated in Figure 6.14, was evolved. The fan is not a part of the gas turbine compressor and may sometimes be mounted at the rear of the engine.

Attempts to increase the efficiency still further lead to even larger fans until they become ducted propellers, or eventually inducted advanced turboprops, so that after many stages of development, we still have to come full circle back to the propeller. Lower bypass engines will still however be required for very high-speed flight.

At this stage, it is essential to note that efficiency is not the only primary consideration in the engine design. The thrust given by the jet engine is almost independent of speed, while the thrust of a propeller, especially if it is of fixed pitch, falls off rapidly both above and below certain speed. It is the thrust that enables an aircraft to fly and dictates its performance; therefore an operator may be more than willing to pay the price provided the thrust requirements are met.

Another difference between the jet and propeller propulsion is that which is related to the fact that the thrust of a jet is almost independent of speed. However, the *power developed by a jet engine*, that is, the (thrust × speed), *varies with the speed*, and there is no satisfactory way of measuring it, either on the ground or in flight. When the aircraft is stationary on the ground, and the engine is running, there is no forward velocity – so that the power is nil, but the thrust may be considerable and can be measured. That is why the performance capability of a jet engine (or a rocket) is given *in terms of thrust and not in terms of power*. Consequently, when an engine drives a propeller, and this applies whether the engine is of the turbine or piston type, the thrust, as we have said, is variable. However, the power produced at the propeller shaft may be considerable even when the aircraft is stationary, and what is more useful is that it can be measured – the propeller acts as a break and is sometimes called brake power – so these engines are compared according to the power they produce and not the thrust, which would be meaningless.

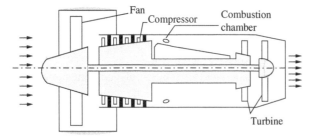

Figure 6.14 Schematic of a turbofan engine.

6.11   The Propeller

The propeller or *airscrew* is the most used of the various system of propulsion in the past. However, at present mostly gas turbine, rather than reciprocating engines, is used for driving propellers. The objective of the propeller is to convert the *torque*, or turning effect, given by the power of the engine, into a straightforward pull or push, called *thrust*.

If a propeller is in front of the engine, it will cause tension in the shaft and so will *pull* the aircraft – such an airscrew is called a *tractor*. If the propeller is behind the engine, it will push the aircraft forward, and it is called a *pusher*.

6.11.1   Working of a Propeller

The cross section of a propeller blade is similar to that of an aerofoil. Therefore, each of the blade is inclined at a small angle (that is, at an angle of attack) to its direction of motion. However, the propeller rotates and moves forward; therefore the direction of the airflow against the blade will be at some angle as in Figures 6.15 and 6.16. This will result in lift and drag on the blade section, just as in the case of an aerofoil. However, in a propeller, we are so much concerned with forces perpendicular and parallel to the airflow, that is, lift and drag, as the force acting along the axis of the aircraft (the thrust force) and at right angle to the rotation (the resistance force). Therefore, its total force on the blade must be resolved into thrust and resistance forces, as shown in Figure 6.16.

The torque acting on the propeller blades will cause a turning moment that opposes the engine torque and also tend to rotate the complete aircraft in the opposite direction to that in which the propeller is revolving. When the propeller is revolving in a steady number of revolutions per minute, then the *propeller torque and the engine torque will be exactly equal and opposite*.

6.11.2   Helix Angle and Blade Angle

The propeller theory is more involved than the aerofoil theory. This is because the local direction of motion of the blade is along a helix rather than a straight line. Also, any section of the propeller

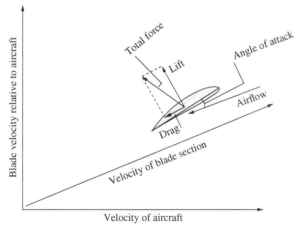

Figure 6.15   Lift and drag acting on a moving propeller blade.

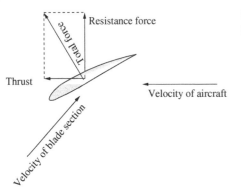

Figure 6.16 Thrust and resistance acting on a moving propeller blade.

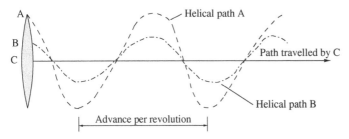

Figure 6.17 Helical path travelled by various sections of propeller blade.

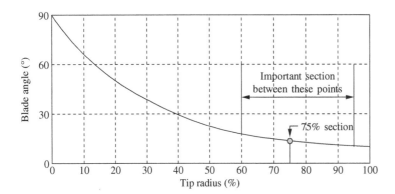

Figure 6.18 Variation of blade angle.

blade travels at a different helix, as illustrated in Figure 6.17. The angle, ϕ, between the resultant direction of the air flow and the plane of rotation, shown in Figure 6.15, is called the *angle of advance* or *helix angle*, and it is different at each section of the blade. The sections near the tip move on a helix of much greater diameter, and they also move at a much greater velocity than those near the boss.

Since all the sections must be set at a small *extra* angle to give the angle of attack, and since for maximum efficiency this extra angle should be approximately the same at all parts of the blade, it is clear that the *blade angle*, or *pitch angle*, must vary in line with the helix angle from the base to tip. Figure 6.18 shows a typical variation of blade angle.

The *blade angle* is the angle that the chord of the propeller section at any particular place makes with the horizontal plane where the propeller is laid flat on its base in this horizontal plane, its axis

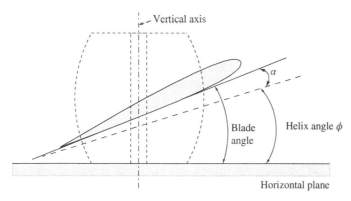

Figure 6.19 Blade angle.

being vertical, as illustrated in Figure 6.19. The figure shows how the blade angle is made up of the helix angle plus the angle of attack.

6.11.3 Advance per Revolution

The blade angle at each section of a propeller is greater than the helix angle, and what is more important is that the distance moved forward in one revolution, termed the *advance per revolution*, is not by any means a fixed quantity, as it depends entirely on the forward speed of the aircraft. For example, if an aircraft is flying at 120 m/s and the propeller is making 1200 rpm, the advance per revolution will be

$$\frac{120}{1200} \times 60 = 6 \text{ m}$$

However, the same aircraft may fly at 100 m/s, with the same revolution of the propeller, and the advance per revolution will be only 5 m; when the engine is run on the ground and there is no forward motion, the advance per revolution will obviously be 0.

For a fixed-pitch propeller, if the angle of a blade section at a radius of r metres is $\theta°$, and if this particular blade section were to move parallel to its chord – that is, with its angle of attack 0° – while at the same time it makes one complete revolution, then the distance travelled forward, x metres, would be a definite quantity and would correspond to the pitch of an ordinary screw. The relation

$$x = 2\pi r \tan \theta$$

will give the distance travelled forward. This can be represented graphically by settling off the blade angle θ from the distance $2\pi r$ drawn horizontally, x being the vertical height, as illustrated in Figure 6.20. If the same operation is carried out at different distances from the axis of the propeller, it will be found that the value of x is practically the same for all sections of the blade, as shown in Figure 6.20, since as the radius r increases there is a corresponding decrease in the blade angle θ and $2\pi r \tan \theta$ remains constant.

6.11.4 Pitch of a Propeller

The forward distance travelled by a propeller per revolution, x, is called the *geometric pitch*, since it depends only on the geometric dimensions and not on the performance of the propeller. The value of the geometric pitch of a fixed-pitch propeller may vary from about 1 m for slow type of aircraft to 5 or 6 m for faster aircraft.

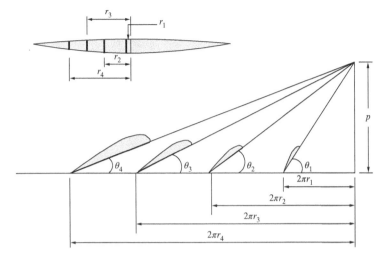

Figure 6.20 Blade angle.

For design considerations, the pitch may be viewed from a different viewpoint. When the advance per revolution reaches a certain value, the thrust becomes zero, the reason being that the angle of attack of each part of the blade has become so small that the aerofoil section of the blade provides no thrust. Note that this corresponds to the small negative angle of attack at which an aerofoil ceases to give lift. The *experimental mean pitch* is defined as *the forward distance travelled by the propeller in one revolution* when it is giving no thrust.

6.11.5 Propeller Efficiency

The *efficiency* of a propeller is the ratio of the useful work delivered by the propeller to the work into it by the engine. We know that the mechanical work done is given by the force multiplied by the distance moved; therefore, when either the force or the distance is zero, the useful work done becomes zero, and the efficiency is zero. Thus when the propeller moves forward in each revolution a distance equal to the experimental pitch, the fact that there is no thrust means that the efficiency is nil. Also, when there is no forward speed, distance moved is zero, and hence there is no work done. Therefore, the efficiency is zero. Between these two extremes are the normal conditions of flight.

Usually a propeller is designed to give the maximum thrust, T, with the minimum torque, Q, that is, to give the *maximum T/Q ratio*. For a high value of T/Q, the value of L/D should be high, and the helix angle should be optimal, which is theoretically around 45°. The high value of L/D is fairly easy and can be achieved with small angle of incidence of the aerofoil relative to the airflow, and this means twisting the blade of the propeller. However, provision of the optimum helix angle is difficult. This requires matching the rotational speed to the forward speed. In practice this is impractical, and the propeller is normally run at nearly constant speed. In any case, the optimum helix angle can only be obtained at one position along the blade, since the blade is twisted. However, the tip of the propeller is moving faster than the inboard sections and thus tends to produce a high proportion of the thrust, so that it is the angle of the tip that is most important. For fixed-pitch propellers, a compromise on the pitch angle has to be made between high efficiency cruising and high thrust for take-off.

For achieving the maximum efficiency, the advance per revolution has to be considerably less than the experimental pitch. The experimental pitch is sometimes called the *ideal pitch*, while the advance per revolution is the *actual pitch*. The difference between the two is called the *slip* and is usually expressed as a percentage.

The efficiency of the propeller can also be expressed as follows. Let V be the forward velocity in m/s, T be the thrust of the propeller in newton, n is the revolutions of the engine crankshaft per second, and Q is the torque exerted by the engine in N-m.

The work done by the thrust generated by the propeller is TV, and the work input to the propeller per revolution is $2\pi Q$. Therefore, the total work input to the propeller is $2\pi nQ$. Thus,

$$\text{Propeller efficiency} = \frac{\text{Work delivered}}{\text{Work put in}}$$
$$= \frac{TV}{2\pi nQ} \times 100\%$$

Example 6.1

The power developed by the engine of an aircraft flying at 75 m/s is 400 kW. If the total drag acting on the aircraft is 5 kN, determine the efficiency of the propeller.

Solution

Given: $P = 400$ kW, $D = 5$ kN, and $V = 75$ m/s.

Therefore,

Work given to the propeller in unit time $= 400 \times 10^3$ J.

Work done by the propeller unit in time $= (5 \times 10^3) \times 75$ J.

The efficiency of the propeller is

$$\eta_{\text{propeller}} = \frac{\text{Work delivered}}{\text{Work put in}} \times 100$$
$$= \frac{(5 \times 10^3) \times 75}{400 \times 10^3} \times 100$$
$$= \boxed{93.75\%}$$

6.11.6 Tip Speed

The power developed by a piston engine depends upon the pressures attained during combustion in the cylinders and on the revolutions of the crankshaft. If the propeller rotates at the same speed as the engine crankshaft, the tip speed of the propeller blades is liable to approach or exceed the speed of sound. This high speed causes compressibility effects, which, in turn, means an increase in torque and decrease in thrust – in other words, a *loss of efficiency*. In the early stages of compressibility, some improvement can be effected by slightly changing the blade angle; if this is done, the loss is not serious as long as the actual speed of the tip does not exceed the speed of sound. In addition to changing the blade angle, a reduction gear is often introduced between the engine crankshaft and the propeller; the reduction is not usually large, perhaps 0.7–1 or 0.8–1, but is just sufficient to reduce the tip speed to a reasonable margin below the speed of sound.

The tip speed of the propeller blade depends on its rotational speed, the forward speed of the aircraft, and the diameter of the propeller. The speed of modern aircraft is such that it is becoming very difficult to keep the tip speed down below the speed of sound. For aircraft speed up to 350 knots (1 knot ≈ 1.85 km/h), the loss of propeller efficiency does not become unacceptably large. However, for aircraft speed more than 430 knots, the loss of efficiency becomes very high and spreads to a larger proportion of the propeller blades so that it affects not only the tips but also the sections of the blade, which should be most efficient. At this stage the only option left is to retire the propeller and switch over to jet propulsion.

Another problem associated with high tip speed is that the noise caused by the propeller is very much intensified, especially in the plane in which the propeller is rotating. This can even cause structural damage to the propeller.

6.11.7 Variable Pitch

For low-speed aircraft, the thrust developed by a fixed-pitch propeller is found to be the greatest when there is no forward speed, that is, when the aircraft is stationary on the ground. The thrust developed under these condition is called the *static thrust*. It is desirable to have a large static thrust since it serves to give the aircraft a good acceleration when starting from rest and thus reduces the take-off run distance required. However, for high-speed aircraft a fixed-pitch propeller designed for maximum speed would have a large pitch and therefore steep pitch angle. Some portion of such blades would strike the air at an angle as high as 70° or more when there is no forward speed, leading to poor static thrust. To overcome this difficulty, variable-pitch propeller has to be employed for high-speed aircraft.

This requirement led to the development of the *constant speed* propeller in which the pitch is automatically adjusted so that the propeller revolves at a given rate decided by the pilot and remains at that rate irrespective of the throttle opening and the manoeuvres of the aircraft. Thus, the engine and propeller can work at high efficiency irrespective of conditions, such as take-off, climb, maximum speed, attitude, and so on.

An extension of the concept of variable pitch leads to a propeller with the pitch variable not only over the range of blade angles that will be required for normal conditions of flight but also beyond these angles in both directions. If the blade can be turned beyond the normal fully coarse position until the chord lies along the direction of flight, thus offering the maximum resistance, the propeller is said to be *feathered*. This condition is very useful on a multi-engined aircraft for reducing the drag of the propeller as a convenient method of stopping the propeller and so preventing it from 'windmilling'. This reduces the risk of further damage to an engine that is already damaged.

The turning of the blade beyond the fully fine position makes the propeller into an effective *air brake*; it has exactly the opposite effect to feathering by causing the maximum drag, which occurs when the blade angle is approximately around 2° or 3°. If the blade angle is further reduced, that is, to negative angles, then instead of allowing the blades to windmill, we can run the engine and produce negative thrust (that is, drag). This produces an excellent brake for use in slowing up the aircraft after landing since it gives high negative thrust at low forward speeds.

6.11.8 Number and Shape of Blades

The propeller must absorb the power given to it by the engine and must have a resisting torque to balance the engine torque; otherwise it will race (that is, there will be a sudden increase in the number of revolutions made by the engine), and both propeller and engine will become inefficient.

For the propeller to meet this condition becomes difficult particularly during climbing since high power is being used at low forward speeds. However, this condition can be satisfied by any one of the methods listed below. It is essential to note that it will be difficult to get efficiency in high-speed flight. Thus, the propeller becomes a compromise like so many things in an aircraft.

The ability of the propeller to absorb power may be increased by increasing the following:

- The blade angle and thus the angle of attack of the blades.
- The length of the blades and thus the diameter of the propeller.
- The rotational speed (rpm) of the propeller.
- The camber of the aerofoil section of which the blade is made.
- The chord of the blades.
- The number of blades.

Even though there are many ways to meet the performance requirement of the propeller, in reality, application of these methods poses considerable challenge. First, the blade angle should be such that the angle of attack is giving maximum efficiency. There is, therefore, little point in trying to absorb more power if, in doing so, the efficiency of the propeller is lost. The second possibility is to increase the diameter, in other words, to increase the blade aspect ratio. This will pose the problem of providing enough ground clearance, in addition to high tip speed. Increase of propeller rpm will lead to high tip speed and consequent loss of efficiency. Increasing camber of an aerofoil would result in increase of the thickness of the aerofoil, which is not a desirable feature since the aerofoil should be thin to avoid loss of efficiency at high speed. Therefore, we are left with only the last two options, namely, increasing the chord (width) of the blades and increasing the number of blades. Both will result in an increase in what is called the *solidity* of the propeller. This really means the ratio between that part of the propeller disc which, when viewed from the front, is solid and the part of which is just air. The greater the solidity, the greater the power absorbed by the propeller.

Increasing the solidity by increasing the chord is *easier* than by increasing the number of blades. However, increasing the solidity by increasing the number of blades is *more efficient*. This is because beyond a limiting length the chord should not be increased, since such an increase would result in poor aspect ratio, leading to blades becoming less efficient. Here, an increase in the number of blades is the most attractive proposition.

6.12 The Slipstream

The propeller produces thrust by forcing the air backwards, and the resultant stream of air that flows over the fuselage, tail units, and other parts of the aircraft is called the *slipstream*. The extent of slipstream may be taken roughly as being that of a cylinder of the same diameter of the propeller. Actually there is a slight contraction of the diameter at a short distance behind the propeller.

The velocity of the slipstream is greater than that at which the aircraft is travelling through the air; the increase in velocity may be as much as 100%, or even more, at the stalling speed of the aircraft. This means that the velocity of the air flowing over all those parts in the slipstream is twice that of the airflow over the other parts, and so the drag is four times as great as the drag of the corresponding parts outside the slipstream. At higher forward speeds the difference is not as great, being only about 50% at normal speeds and as small as 10% at high speeds. The extra velocity of the slipstream may be beneficial in providing more effective control for rudder and elevators, especially when the aircraft is travelling slowly through the air, for example, when taxiing, or taking off, or flying near the stalling speed.

In addition to increased velocity, the propeller imparts a *rotary motion* to the slipstream in the same direction as its own rotation; so it will strike one side only of such surface as the fin and so may have considerable effects that are compensated for in normal flight. For example, by offsetting the fin so that it does not lie directly fore and aft, the balance then will be upset when the engine stops and the slipstream ceases to exert its influence.

6.13   Gyroscopic Effect

The rotation of the propeller in a piston engine or the compressor in the case of a jet engine may cause a slight *gyroscopic effect*. A rotating body tends to resist any change in its plane of rotation, and if such change does take place, there is superimposed a tendency of rotation. Thus if the propeller rotates clockwise when viewed from the cockpit, the nose will tend to drop on a right-hand turn, and the tail to drop on a left-hand turn.

6.14   Swing on Take-Off

Swing is a tendency to turn to nose side during take-off, owing to some asymmetry of the aircraft. The following are the main causes for swing:

- The pilot himself is not seated symmetrical in the cockpit.
- The asymmetric effects of the propeller. If the propeller rotates clockwise, the *torque reaction* will be counterclockwise, the left-hand wheel will be pressed on the ground, and the extra friction should tend to yaw the aircraft to the left. However, the torque reaction may be compensated; in that case, the behaviour of the aircraft will depend on how it is compensated.
- For a clockwise propeller, the *slipstream* also will rotate clockwise and will probably strike the fin and rudder on the left-hand side, again tending to yaw the aircraft to the left. However, the slipstream may also be compensated.
- The gyroscopic effect will come only when the tail is being raised. Again the tendency will be to swing to the left if the propeller rotates clockwise.

The swing effect due to propeller rotation can largely be eliminated by the use of propeller rotating in opposite direction on multi-engined aircraft by contra-rotating propellers on single-engined aircraft and by jet propulsion or rocket propulsion instead of propellers. Contra-rotating propellers give greater blade area, or solidity, that is required to absorb large power. Also they almost eliminate all the asymmetrical effects of slipstream, propeller torque, and gyroscopic action. Indeed, the pilot usually does not realise the existence of these asymmetrical effects – until he loose them.

6.15   Thermodynamic Cycles of Jet Propulsion

Gas turbines are based on the *Brayton cycle*. All jet engines and gas turbines are heat engines that convert thermal energy into useful work. The useful work may be in the form of mechanical power, as from a shaft that may be used to drive a propeller, a vehicle, a pump, an electric generator, or any other mechanical device. In jet engine applications, the work is in producing compressed air and combustion products that are then accelerated to provide reaction propulsion.

6.15.1   Efficiency

Thermal efficiency, η_{th}; propulsive efficiency, η_{Pr}; and combustion efficiency, η_{Co}, are the three important parameters used to assess the engine performance. The definition and the typical range of these parameters are listed in Table 6.1.

Table 6.1 Typical range of thermal, propulsive, and combustion efficiency.

η_{th}	$\dfrac{\text{Work output}}{\text{Input thermal energy}}$	5–50%
η_{Pr}	$\dfrac{\text{Propulsive work delivered}}{\text{Mechanical energy produced by the engine}}$	5–40%
η_{Co}	$\dfrac{\text{Output thermal energy}}{\text{Input chemical energy}}$	90–99%

Figure 6.21 T–s diagram of Brayton cycle.

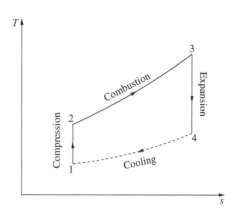

6.15.2 Brayton Cycle

The Brayton cycle is the fundamental constant pressure gas heating cycle used by gas turbines. It consists of isentropic compression, constant pressure heating, isentropic expansion, and constant pressure cooling (absent in open cycle gas turbines), as illustrated in the T–s diagram in Figure 6.21.

6.15.3 Ramjet Cycle

A ramjet uses the open Brayton cycle. In the diagram of supersonic ramjet engine shown in Figure 6.22, a two-dimensional supersonic intake is shown downstream of which is a divergent

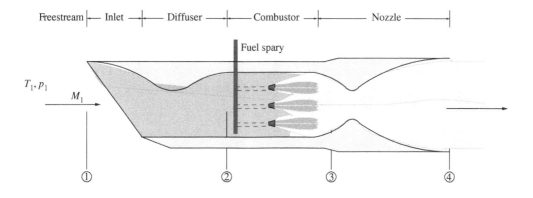

Figure 6.22 Schematic diagram of supersonic ramjet engine. (1) Freestream; (2) fuel spray; (3) nozzle entry; (4) nozzle exit.

subsonic diffuser. Fuel is then injected into the compressed air and evaporates producing a mixture that is ignited when it reaches the flame front. The flame holders provide the turbulent circulation necessary to stabilise the flame, since deflagration velocities are usually much smaller (< 10 m/s) than the average velocity of air in the combustor. The combustion products are then exhausted through the nozzle.

To understand how thrust is produced if we assume that the flow of fuel is negligible compared with the air mass, then the exhaust flow will be at approximately the same Mach number as the input flow. However, the total temperature of the exhaust is much higher, and the exit velocity will be correspondingly higher than the input velocity. This difference in velocity (and momentum) produces thrust.

The temperature rise in the intake diffuser is related to the freestream Mach number M_1 as

$$\frac{T_2}{T_1} = 1 + \frac{\gamma - 1}{2}M_1^2$$

The efficiency in terms of the freestream Mach number M_1 is

$$\eta = \frac{\dfrac{\gamma - 1}{2}M_1^2}{1 + \dfrac{\gamma - 1}{2}M_1^2}$$

where $\gamma = c_p/c_v = 1.4$ is the ratio of specific heats of air.

Maximum efficiency is reached if temperature rise in the combustor is small. Ramjets are inefficient at subsonic speeds, and their efficiency improves at supersonic speeds.

At hypersonic speeds, the compression and dissociation processes make full diffusion unattractive, and supersonic combustion is being researched. A scramjet slows the air down to low supersonic speeds and then burns high flame velocity fuels such as hydrogen or methane.

6.15.4 Turbojet Cycle

Adding a compressor to a ramjet powered by a turbine in the exhaust allows increased combustor inlet temperature and a consequent increase in possible thermal efficiency. The turbine however is limited in the temperature it can handle, so maximum power is also limited.

The $T–s$ diagram of turbojet cycle is shown in Figure 6.23. As shown in the $T–s$ diagram, the presence of the compressor allows us to raise the combustor inlet temperature. The raising of the combustor segment increases the cycle area and the thermal efficiency.

The $T–s$ diagrams of turbojet cycle without afterburner and with afterburner are shown in Figures 6.23 and 6.24, respectively. The addition of an afterburner (5–6) allows thrust augmentation, as seen from the increased area of the diagram shown below. The afterburner operates in the higher entropy range and has lower efficiency than the base turbojet.

6.15.5 Turbofan Cycle

A turbofan diverts some of the pressure energy of the core flow to power a fan that moves a larger mass flow, providing an increase in thrust and propulsive efficiency. Turbofans normally have two or three shafts. Since the diameter of the fan is larger, the same tip speed can be achieved at a lower rpm than the smaller-diameter compressor, and two shafts become necessary. The alternate method is to employ a gearbox to step down the shaft speed that is used in some smaller turbofans. In most turbofans, however a multistage low-pressure (LP) turbine is used to extract

Figure 6.23 *T–s* diagram of turbojet cycle without afterburner.

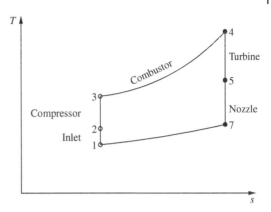

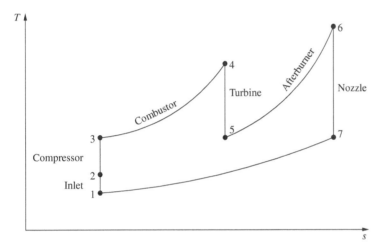

Figure 6.24 *T–s* diagram of turbojet cycle with afterburner.

the same energy with smaller stage loadings and lower tangential velocity. The smaller-diameter high-pressure (HP) compressor is run with one or two turbine stages with higher tangential velocity than the LP turbine.

6.16 Summary

Thrust is the force that opposes drag and enables the airplane going forward. In a steady level flight, the thrust must be equal to the drag. Thrust generated by an engine relies on the reaction of the gas ejected through the engine.

The thermal energy produced by burning the fuel and air is converted into the mechanical work done in propelling the aircraft against the drag.

All jet engines, which are also called *gas turbines*, work on the same principle. The major parts of a jet engine are fan, compressor, combustion chamber, turbine, mixer, and nozzle.

In 1903, the Wright Brothers flew The Flyer with a 12-horse power gas-powered engine. From 1903, the year of the Wright Brothers first flight, to the late 1930s, the gas-powered reciprocating internal- combustion engine with a propeller was the sole means used to propel aircraft. It was

Frank Whittle, a British pilot, who designed and patented the first turbojet engine in 1930. The Whittle engine first flew successfully in May 1941.

Jet engines can be broadly classed as turbojets, turboprops, turbofans, turboshafts, and ramjets. The turbojet engine is a reaction engine.

Turboprop engine is a jet engine attached to a propeller.

A turbofan engine has a large fan at the front, which sucks in air. Most of the air flows around the outside of the engine, making it quieter and giving more thrust at low speeds.

Turboshaft is another form of gas turbine engine that operates much like a turboprop system. It does not drive a propeller. Instead, it provides power for a helicopter rotor.

The ramjet is the simplest jet engine and has no moving parts. The speed of the jet 'rams' or forces air into the engine. It is essentially a turbojet in which rotating machinery has been omitted.

The turbojet, usually referred to as the *jet engine*, is the commonly used aircraft engine at present. In principle, the turbojet is a very simple form of propulsion unit based on the *gas turbine engine*.

The turbojet has no reciprocating parts, such as the piston-cylinder in the piston engine.

The construction of turboprop engine is almost the same as the turbojet engine, but more of the available energy in the exhaust is used to drive the turbine.

The ramjet is used for propelling missiles and early stages of spacecraft flight, although not used for aircraft propulsion.

The reaction caused by the momentum of the combustion products leaving the nozzle is the thrust generated by the engine. The thrust generated can be expressed as

$$\text{Th} = \dot{m}\, u_e + \left(p_e - p_\infty \right) A_e$$

Ideal ramjet is essentially a hypothetical engine with simplified cycles involving isentropic processes.

Rocket propulsion is straightforward in principle. In a rocket motor, the fuel and oxidiser are stored in the motor itself. Thus, a rocket is essentially a non-air-breathing engine unlike a turbine engine that takes in air from the atmosphere as the oxidiser.

Rocket motors come in two basic types, namely, solid fuel type and liquid propellant type. In solid fuel type, the fuel and oxidiser are combined in a stable solid form, as in a simple firework. In a liquid propellant type, two chemicals, one usually the fuel and the other the oxidiser, are mixed together and burned in a combustion chamber. Rocket motors have a very high thrust-to-weight ratio.

Apart from the simple chemically fuelled rockets, several more advanced types of rockets have been developed. These include the rockets in which ionised particles or plasma is accelerated to very high speeds by electrostatic or electromagnetic forces. Some of these have been used on spacecraft, normally as low-thrust control jets.

Propeller engines are essentially internal combustion engines. In a propeller engine, the engine that drives the propeller may itself be a gas turbine, and in this case the designer has the choice to allot the proportion of the power delivered by the propeller and jet as per the design considering as in the *turboprop system*. The thrust generated is

$$T = \dot{m}\, V$$

A jet engine generates thrust by giving a large increase in speed to a relatively smaller mass than a piston or propeller engine. Therefore, to make a jet engine more efficient, the design should be such that a larger mass of air is somehow given a smaller increase in speed. The method used is to increase the size of the compressor fan to allow a portion of the air to pass around the outside of the engine.

The propeller or *airscrew* is the most used of the various system of propulsion in the past. However, at present mostly gas turbine, rather than reciprocating engines, is used for driving propellers.

The propeller theory is more involved than the aerofoil theory. This is because the local direction of motion of the blade is along a helix rather than a straight line. Also, any section of the propeller blade travels at a different helix.

The *blade angle* is the angle that the chord of the propeller section at any particular place makes with the horizontal plane where the propeller is laid flat on its base in this horizontal plane, its axis being vertical.

The *efficiency* of a propeller is the ratio of the useful work delivered by the propeller to the work into it by the engine.

The power developed by a piston engine depends upon the pressures attained during combustion in the cylinders and on the revolutions of the crankshaft.

The rotation of the propeller in a piston engine or the compressor in the case of a jet engine may cause a slight *gyroscopic effect*.

Swing is a tendency to turn to nose side during take-off, owing to some asymmetry of the aircraft.

Gas turbines are based on the *Brayton cycle*.

Thermal efficiency, η_{th}; propulsive efficiency, η_{Pr}; and combustion efficiency, η_{Co}, are the three important parameters used to assess the engine performance.

The Brayton cycle is the fundamental constant pressure gas heating cycle used by gas turbines. It consists of isentropic compression, constant pressure heating, isentropic expansion, and constant pressure cooling.

A ramjet uses the open Brayton cycle.

7

Level Flight

7.1 Introduction

The flight of an aircraft usually consists of the take-off, climb, steady flight at a constant height, the approach, and landing. In a steady level flight, the *lift* will be acting vertically upwards since the direction of flight is horizontal. The *weight* of the aircraft acting downwards is balanced by the lift. The lift can be generated only when the aircraft is moved forward, and for this the *thrust* required is provided by the propeller or jet. The forward motion of the aircraft will be opposed by the *drag*.

7.2 The Forces in Level Flight

In a steady level flight, the forces acting on an aircraft are the lift, weight, thrust, and drag, as shown in Figure 7.1. These four forces keep the aircraft in its state of steady level flight. The lift, L, acts vertically upwards through the *centre of pressure* (*cp*). The weight, W, of the aircraft acts vertically downwards through the centre of gravity (*cg*). The thrust of the engine, T, acts horizontally in the forward direction. Finally, the drag, D, acts horizontally backwards.

It is essential to note that for simplifying the analysis, all the weight is considered as acting through one point, namely, the *cg*, and all the lift as acting at the centre of pressure, so that we may assume that the resultant of all the drag acts at one point as shown in Figure 7.1. However, the actual position of the drag depends on the relative resistance of different parts of the aircraft.

7.3 Equilibrium Condition

Now let us examine the conditions under which the L, W, T, and D balance the aircraft and keep it travelling at a constant altitude (that is, height) at uniform velocity in a fixed direction, referred to as *equilibrium*. Thus for a state of equilibrium in a straight and level flight, the forces acting on the aircraft must be balanced. In other words, for equilibrium, the following needs to be considered:

- The lift, L, should be equal to the weight, W. This condition of $L = W$ will keep the aircraft at a constant altitude.
- The thrust, T, should be equal to the drag, D. This condition of $T = D$ will keep the aircraft moving steadily at a constant speed.

In addition to the above two conditions, a third condition is also required for the equilibrium of aircraft in steady straight and level flight. To maintain the straight and level flight, the aircraft

Introduction to Aerospace Engineering: Basic Principles of Flight, First Edition. Ethirajan Rathakrishnan.
© 2021 John Wiley & Sons, Inc. Published 2021 by John Wiley & Sons, Inc.
Companion Website: www.wiley.com/go/Rathakrishnan/IntroductiontoAerospaceEngineering

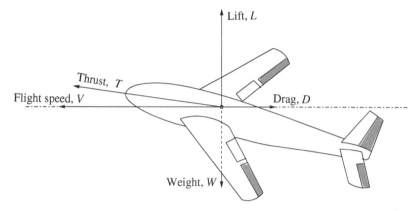

Figure 7.1 Forces acting on an aircraft in steady level flight.

should not be allowed to rotate. The prevention of the aircraft from rotating depends not only on the magnitude of the four forces but also on the position at which they act. If the centre of pressure (*cp*) is behind the centre of gravity (*cg*), the nose of the aircraft will tend to drop, and the tail will tend to rise, and vice versa if the *cp* is ahead of *cg*. Any difference in the line of action of the thrust and drag also can make the aircraft to rotate. If the line of thrust is above the line of drag, these forces also will make the nose to drop. Such tendencies could be prevented by the pilot using the control surfaces of the aircraft.

7.4 Balancing the Forces

The lift that acts through the *cp* will depend on the position of the wings. Therefore, in the design, attention must be paid to place the wings in correct position along the fuselage. However, making the lift to act at the *cp* becomes complicated due to the fact that a change in the angle of attack causes the lift to move. If the angle of attack is increased, the pitching moment about the *cg* will become more nose-up and tend to increase the angle even further. The acting of the weight through the *cg* depends on the weight and position of every individual part of the aircraft and the loads that they carry. However, during a flight, the *cg* can move due to the consumption of fuel, dropping of bombs, or movement of passengers.

The line of action of the thrust is dictated by the position of the propeller shaft or centreline of the jet from the engine nozzle, which in turn depends on the position of the engines.

Finally, in the line of action of the drag, the total drag is made up of the drag of all the separate parts. The designer either should estimate the drag of each part separately and find the total drag and its line of action or should rely on wind tunnel experiments on a model or computed values. The line of action of the drag also would change at different angles of attack.

For steady flight along a straight line, it is essential to balance the four forces so that they produce no resultant force. Also, their line of action must also be such that they produce no resultant moment, causing the aircraft to rotate either nose-up or nose-down. When there is no resultant moment M, the aircraft is said to be *trimmed*. That is, the condition for trim is $M = 0$.

For trimmed flight, all the four forces should act through a single point, as illustrated in Figure 7.2. However, this is not generally practical, as there are many factors that tend to alter the line of action. For example, lowering the undercarriage tends to shift the line of resultant

Figure 7.2 An aircraft in flight with forces acting through a single point.

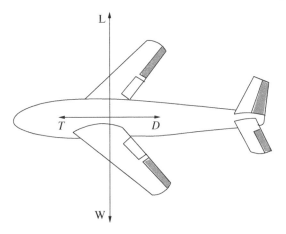

Figure 7.3 Moment due to lift and weight balanced by moment due to drag and thrust.

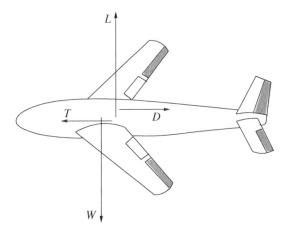

drag down. However, it is possible to balance the moment caused by the drag and thrust being out of line, by arranging the lift and weight to be out of line by a distance that causes exactly the necessary balancing moment, as shown in Figure 7.3.

Consequently, the line of action of forces tends to change according to the altitude at which the aircraft flies, fuel weight, etc.. Hence, there is no simple design solution to ensure that the resultant moment will always be zero. Therefore, to keep the aircraft trimmed, the pilot should use control surface to adjust the centre of moment causing the imbalance.

7.4.1 Control Surface

The aircraft can be trimmed to give no resultant moment with an auxiliary lifting surface called a tailplane. The lift on the tailplane can be regulated by the pilot, and this adjusts the moment that it applies. The tailplane can produce lift in either the +ve (upward) or −ve (downward) direction, as shown in Figure 7.4, to produce the required moment for trim.

To change the tailplane lift, either the whole surface can be pivoted, or the rear part of the tail surface, termed *elevator*, can be hinged up or down. In practice, for small adjustment to the trim, it is a common practice to provide a very small hinged surface or *trim tab* in addition to the main elevator.

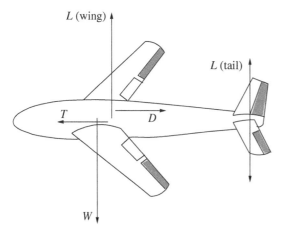

Figure 7.4 Lift generated by the tailplane.

7.4.2 Tail-Less and Tail-First Aircraft

Tail-less aircraft are essentially flying machines with a large degree of sweepback or even delta-shaped wings. Although this type of aircraft may appear to have no tail, the exact equivalent of tail is found at its wing tips, the wings being, in fact, swept back so that the tip portion can fulfil the function of the tailplane in the conventional aircraft. Thus, in essence, the tail-less type has two tails instead of one.

There are some designs in which the tail of the aircraft is located ahead of the wing. This type is termed tail-first or *canard* configuration. The tail in front can hardly be called a tail, and this surface is commonly known as the fore plane.

7.4.3 Forces on Tailplane

When the lift, weight, thrust, and drag are balancing themselves, as in a steady level flight, the tailplane just acts as a standby. Therefore, usually the tailplane will be set at such an angle that for the aircraft in cruise the tail will be at zero angle of attack, thereby producing no lift. At flight speeds higher than cruise speed, the lift coefficient of the tail must compensate for the higher dynamic pressure; otherwise the lift would be greater than the weight. This means that the aircraft must be trimmed a little more nose-down. In doing so, the centre of lift of the wing will move back, giving a nose-down pitching moment. However, if the angle of attack of the tailplane was zero for cruise, then it will become −ve, and the tail will generate a downforce, as illustrated in Figure 7.5, producing a nose-up pitching moment that will counter the nose-down moment caused by the wing.

As the flight speed increases, the pitch normally has to make a small nose-down trim adjustment. At low speed, the nose of the aircraft must be raised to increase the angle of attack. This means that the tail lift becomes +ve, as shown in Figure 7.6. As the tailplane is required to produce upward and downward force, it is usually a symmetrical profile and therefore produces no lift at zero angle of attack. On a tail-first or canard aircraft, the fore plane is set at a slightly higher angle of attack than the wings for reasons of stability, and both wings produce lift in normal flight.

7.4.4 Effect of Downwash

In many types of aircraft, the air that strikes the tailplane has already passed over the main wing, and the trailing vortices from there will cause a downwash on to the tailplane, as illustrated in Figure 7.7.

Figure 7.5 Down load from tail balancing the effect of rearward location of lift.

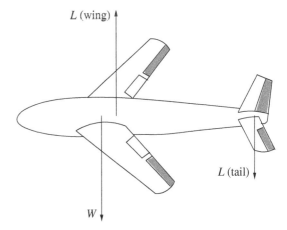

Figure 7.6 Up load from tail balancing the nose-up moment caused by the wing.

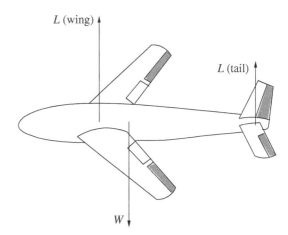

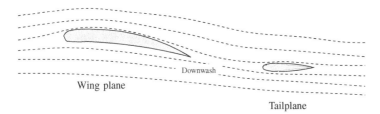

Figure 7.7 Effect of downwash on the tail.

The angle of this downwash may be at least half the angle of attack on the main wing. Thus, the angle of incidence on the tailplane will be $\alpha/2$ when the main wing is at α. Therefore, if the tailplane were given an angle (the angle of incidence is the angle at which the aircraft wing is attached to the aircraft fuselage) of incidence of $\alpha/2$, the air stream would approach the tail head-on, and a symmetrical tail will provide no force upwards or downwards. However, the angle of downwash will change with the angle of attack of the main wing, and this makes it impossible to decide on a specific angle at which the tailplane has to be set.

The setting of the wing will also affect the stability of the aircraft, and further difficulties arise from the fact that in a propeller-driven aircraft, the tailplane is usually in the slipstream, which is a rotating mass of air and will therefore strike the two sides of the tailplane at different angles. In jet-driven aircraft, the tailplane is often set very high, to keep it clean of the hot jets, and this may cause trouble since it may be shielded by the main planes at large angles of attack, resulting in what is called a *deep stall.*

7.4.5 Varying the Tailplane Lift

Usually the elevator hinged at the rear of the tailplane is used to alter the tailplane lift. Moving the elevator up or down alters the camber of the tail surface and thus changes the lift. However, when the flight speed is supersonic, changes to camber do not produce much of change in lift, and it is better to move the whole tail surface as single slab. This moving slab tailplane is sometimes used on low-speed aircraft also. On large transport aircraft, the two approaches may be combined. The whole tailplane can be moved, mainly to provide for trimming, and a hinged elevator is also used mainly for control and manoeuvre.

7.4.6 Straight and Level Flight

In straight and level flight, the lift balances the weight and thrust is equal to drag. When there is no load on the tailplane, the conditions for balance are the following:

- $L = W$.
- $T = D$.
- The 'nose-down' pitching moment of L and W must balance the 'tail-down' pitching moment of T and D. The forces L and W are equal and opposite parallel forces. Therefore, the moment caused by L and W is given by one of the forces multiplied by the perpendicular distance between them. Therefore, if x is the distance between L and W, the moment is Lx or Wx Newton-metre. Similarly, if y is the distance between T and D, the moment due to the T and D is given by Ty and Dy Newton-metre.

Therefore, the third condition is that

$$Lx = Ty$$

At this stage, it will be useful to note that a wing of an aircraft generates a large amount of lift and very small amount of drag, and all other parts generate only drag and contribute nothing to the lift. Therefore, it is essential that the wings provide all the lift; however by intelligent design, even parts such as the fuselage may be made to generate some lift. However, even the total increase of lift from such a source will be small, whereas the addition of the parasite drag of fuselage, tail, and undercarriage will form a large portion of the total drag. In old design this drag addition will be as much or even more than the drag of the wing, but in modern aircraft with 'clean' lines, the proportion of parasite drag has been greatly reduced.

Example 7.1

An aircraft in level flight experiences a drag of 1.3 kN. The *cp* is 25 mm behind the *cg*, and the line of drag is 150 mm above the line of thrust. If the weight of the aircraft is 12 kN, find the load on the tailplane, which is 6.22 m behind the *cg* required to maintain the flight.

Solution

Given: $D = 1300$ N and $W = 12\,000$ N.

The forces acting on the aircraft are as shown in Figure 7.8.

Figure 7.8 The forces acing on the aircraft in level flight.

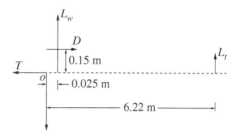

The total lift is

$$L = L_w + L_t$$

where L_w and L_t are the lift of the wing and tail, respectively.
In level flight, $L = W$; therefore,

$$L_w + L_t = 12\ 000$$

Also,

$$T = D = 1300$$

Taking moments about o, we have

$$L_t \times 6.22 + L_w \times 0.025 = D \times 0.15$$

Therefore,

$$L_t \times 6.22 + (L - L_t) \times 0.025 = D \times 0.15$$
$$L_t(6.22 - 0.025) + L \times 0.025 = D \times 0.15$$
$$L_t = \frac{D \times 0.15 - L \times 0.025}{6.195}$$
$$= \frac{1300 \times 0.15 - 12\ 000 \times 0.025}{6.195}$$
$$= \frac{195 - 300}{6.195}$$
$$= -\frac{105}{6.195}$$
$$= \boxed{-16.95\ \text{N}}$$

The −ve sign implies that L_t is acting downwards. Thus a down lift of 16.95 N has to be generated by the tailplane.

7.4.7 Relation Between Flight Speed and Angle of Attack

The flight speed range of an aircraft can vary from the maximum speed that can be attained to the minimum speed at which the aircraft can be kept in the air, both without losing height. This *speed range* is very wide for modern aircraft: the maximum speed may be 1000 knots (1 knot is approximately about 1.85 km/h) or even more, and the minimum speed (with flaps lowered) less than 15 knots.

An aircraft flying in level flight at different speeds will be flying at different angles of attack. Level fight can be at different altitudes, that is, at different heights with reference to the ground. For every air speed, there is a corresponding angle of attack at which level flight can be maintained (of course, provided the weight of the aircraft does not change).

In level flight as we know the lift is equal to the weight, that is,

$$L = W = \frac{1}{2}\rho V^2 S C_L$$

Assuming the weight remains constant, we have the condition that the lift remains constant and equal to the weight. Also, the wing area S is a constant. Further, $\frac{1}{2}\rho V^2$ represents the dynamic pressure, which is the difference between the pitot and static pressures. This in other words represents the *indicated air speed* (IAS). Therefore, when the IAS increases, the lift coefficient, C_L, must be reduced to maintain constant lift. Similarly if $\frac{1}{2}\rho V^2$ goes down, C_L must go up to maintain constant lift. The lift coefficient depends on the angle of attack of the wings; the greater the angle of attack (up to the stalling angle), the greater the value of C_L.

7.5 Range Maximum

Maximising the range or the flying distance of an aircraft is an important requirement. The range of an aircraft is dictated by the engine, propeller, and the wind condition at the flying altitude. The fuel is burnt in an engine to achieve the maximum possible thermal energy and convert the energy into mechanical work. Therefore, to get the maximum amount of work for a given amount of fuel, the combustion of the fuel should be with the highest possible efficiency, releasing the maximum amount of thermal energy, and then the energy must be converted to mechanical work in the most efficient way. To get the most heat from fuel, it must be properly burned. For proper burning, the mixture of air and fuel must be optimum. In a piston engine, what is referred to as 'weak or lean mixture' is approximately the correct mixture to burn the fuel efficiently. What is meant by a *rich mixture* is the mixture in which the fuel will not burn properly and release lesser energy than that can be obtained with an efficient burning when the mixture is weak.

The problem of converting the calorific value of the fuel by burning it with the oxygen in the air and then converting into mechanical work is governed by the law of thermodynamics. Even in the best engine, the efficiency of conversion of thermal energy to mechanical energy is just about 30%.

7.5.1 Flying with Minimum Drag

The lift-to-drag ratio will be a maximum when the drag is a minimum. The L/D ratio for the wing alone will be always greater than the L/D ratio of the entire aircraft. This is because the lift of an aircraft is almost entirely due to the wing alone, and the lift generated by other parts of the aircraft is insignificant. However, the drag due to the fuselage, nacelle, etc. is also considerable, and hence this constitutes a significant portion of the total drag.

The angle of attack that gives the best L/D ratio will be the same at all height or flying altitude and at all weight of the aircraft. Therefore, drag minimum simply implies the flight at the *best attitude* and has nothing to do with the density of air, or the payload. This implies that the IAS (that is, $\frac{1}{2}\rho V^2$) will be the same, whatever be the altitude, but will increase slightly for increased load. The same IAS means the same drag at any height and therefore the same range.

On the other hand, the higher the speed that must be used for increased weight means greater drag, because for the same L/D ratio, if the lift is greater, the drag also becomes higher to maintain the constant value of L/D.

Thus, to obtain the maximum range, the aircraft must fly at a given angle of attack, that is, at a given IAS, it may fly at any height but should carry a minimum load. If extra payload is to be carried, the air speed (that is, the flight speed) must be increased since in level flight $L = W$, the increase of

W can be accompanied only through the increase of velocity in the lift expression. $L = \frac{1}{2}\rho V^2 S C_L$, since other than V all the parameters in L are invariant at a given altitude and given angle of attack.

7.6 Altitude Effect on Propeller Efficiency

From our discussions on level flight, it is obvious that aircraft will get the same range as long as it flies at the same IAS, whatever be the height. Now, although the drag is the same at the same IAS at all heights, the power is not. Though appears strange, this is an important fact. If it were not so, for a given power, the aircraft will experience higher IAS with the increase of altitude. Therefore, it would be an advantage to fly high because for the same power, the higher the flying altitude the greater the true speed. However, this is not the case in practice. This is because of the following reason. Power is the rate of doing work. A given amount of fuel gives only a specified amount of power (Newton-metre); however long the aircraft has to fly. Even so, if the work has to be done quickly, with the aim of maintaining the generation of certain amount of *thrust* through a certain *distance* in a certain *time*, then the power will depend on the thrust and the distance and the time. In other words, for this aim, the power depends on the *thrust* and *velocity*. Now we may have the doubt that this velocity is indicated velocity or true velocity. This can easily be clarified if we note the simple fact that the distance to be covered (that is, the range) is the true or actual distance. Therefore, the speed that dictates the power requirement is the true air speed. That is, the larger the distance to be covered, the greater the true air speed for the same IAS, and therefore the greater the power required, although the thrust and drag remain the same.

We know that the reciprocating engine can be designed to work efficiently at some considerable height above the sea level. Therefore, at sea-level altitude, if the aircraft is flying at the best speed for range, the thrust will be a minimum, which is desirable. However, owing to the lower speed, little power will be required for the engine. This sounds satisfactory, but actually it is not economical. The engine must be throttled, the Venturi tube in its carburettor is partially closed, the engine is held in check and does not run at its designed power and, what is more important, does not give its best efficiency. For maximum range, both aircraft and engine should be used to the best advantage, and this can easily be done if the flying height (altitude) is such that the aircraft will fly at the lowest IAS and the engine will be operating at the best efficiency, that is, the engine is operated with its throttle valve fully open, but still flies with weak fuel–air mixture.

At this stage, it is essential to understand what happens at greater heights. At the same IAS, more power is required when the altitude increases. However, if the throttle is fully open, we cannot get more power without using a richer mixture. Therefore, either the speed should be reduced, or mixture must be enriched. Consequently, both will result only in uneconomical operation. Thus, the best performance is achieved at a specified height, but the height is determined by the engine efficiency (that is, by propulsive efficiency) and not by the aircraft. Indeed an aircraft would be equally good at all heights. The best height is not usually very critical, nor is there generally any great loss in range by flying below that height.

7.7 Wind Effect on Range

If a flight is from place 1 to place 2 and back to place 2, the wind of any speed and any direction will adversely affect the range of action. The wind usually changes its direction and increases its velocity with height. Even in straight and level flight, the head wind (wind opposing the flight direction) or tail wind (wind in the direction of flight) can influence the range significantly. For example, the

Trans Atlantic flight from London to New York by, say, 747 aircraft will take about 6 hours when there is no head or tail wind. This flight time can decrease even of the order of 1 hour when there is a strong tail wind and may become 7 hours when there is a strong head wind.

7.8 Endurance of Flight

Endurance is the time duration for which an aircraft can stay in the air for a given quantity of fuel. To get maximum endurance, the least possible fuel should be used in a given time, that is, the power used should be minimum to get maximum endurance. We know that the power, P, is given by

$$P = D\,V$$

where D is the drag and V is the true air speed. Therefore, from the power point of view, slower speed for a given power implies increased drag. Therefore, a proper adjustment between speed, V, and drag, D, is essential for adjusting the endurance and the range. However, the speed for the best endurance is less than the speed for the best range. Another important aspect in this situation is that the speed of concern in the range and endurance is the true air speed; thus the lower the height, the better the endurance.

The optimum speed for maximum endurance is usually uncomfortably low for accurate flying, even the speed for the best range is not always easy. However, as neither the maximum endurance nor the maximum range is critical, the pilot is often recommended to fly at a somewhat higher speed. As we saw, maximum value of $C_L^{3/2}/C_D$ is the requirement for maximum power, ensuring maximum endurance.

Therefore, for maximum endurance, the aircraft should fly at an angle of attack that gives the best value of $C_L^{3/2}/C_D$. The best value of $C_L^{3/2}/C_D$ for an aircraft will be at a greater angle of attack and a lower speed than for range.

7.9 Range Maximum

Range of an aircraft is the distance it can cover in flight for a given quantity of fuel. Therefore for an aircraft, flying for maximum range means flying with minimum drag. In other words, the aircraft fly with maximum efficiency no matter it uses propeller engine or jet engine. Even so, if, when we fly with minimum drag, either the propulsive system or the engine, or both, is very inefficient, therefore it is essential to make some compromise probably by flying rather faster than the minimum drag speed.

An aircraft will fly the same range, irrespective of the height at which it flies, provided it flies with minimum drag condition (that is, minimum drag speed). In the propeller-driven aircraft, it is beneficial to fly faster than the minimum drag speed because, by doing so, engine and propeller efficiency is improved. Also, it is advantageous to fly at a certain height at which the engine–propeller combination is more efficient. However, in the case of aircraft propelled with jet engine, this is not true. The following are the important reasons for this difference between the propeller and jet aircraft:

- The thrust of a propeller engine decreases as the flight speed increases, whereas the thrust of a jet engine is nearly constant at all speeds (at the same rpm).
- The fuel consumption of a reciprocating (propeller) engine is approximately proportional to the power developed, but for a jet engine, the fuel consumption is approximately proportional to the thrust.

The flying range of an aircraft will increase if it is flown faster than the minimum drag speed, but the drag will be slightly higher. The thrust, being equal to the drag, will also be slightly higher. This will lead to higher fuel consumption. The speed, on the other hand, will be considerably greater, and so the range of the aircraft will increase considerably. Thus, essentially the range of an aircraft depends on getting maximum of speed compared with thrust or speed compared with drag. That is, for maximum range, the aircraft must fly at minimum drag speed. Therefore, to get maximum range, jet aircraft must fly faster than propeller aircraft – the difference being due to the different relationship between the efficiency and speed in the two systems.

For an aircraft flying at a constant altitude, the drag and thrust will be equal, if it flies at the same IAS, but the true air speed will be greater. Even so, the overall efficiency of the aircraft is dictated by the true air speed; thus when the true air speed is increased, the efficiency will go up. For the engine for which the fuel consumption remains proportional to thrust, the range will be slightly increased. Therefore, to get higher range, a jet aircraft should fly at higher altitude.

7.10 Endurance of Jet Engine

For jet engines, fuel consumption is approximately proportional to thrust. Therefore, for maximum endurance, it should fly with minimum thrust, that is, with minimum drag. Therefore, the endurance speed of a jet aircraft corresponds closely to the range speed of a propeller aircraft. The thrust, and hence the fuel consumption, should be the same at the same IAS at any height; the endurance of a jet aircraft is not affected by the altitude at which it flies.

7.11 Summary

The flight of an aircraft usually consists of the take-off, climb, steady flight at a constant height, the approach, and landing.

In a steady level flight, the forces acting on an aircraft are the lift, weight, thrust, and drag.

For equilibrium, the lift, L, should be equal to the weight, W. This condition of $L = W$ will keep the aircraft at a constant altitude. The thrust, T, should be equal to the drag, D. This condition of $T = D$ will keep the aircraft moving steadily at a constant speed.

For steady flight along a straight line, it is essential to balance the four forces so that they produce no resultant force.

For trimmed flight, all the four forces should act through a single point.

The aircraft can be trimmed to give no resultant moment with an auxiliary lifting surface called a tailplane. The rear part of the tail surface, termed *elevator*, can be hinged up or down. In practice, for small adjustment to the trim, it is a common practice to provide a very small hinged surface or *trim tab* in addition to the main elevator.

Tail-less aircraft are essentially flying machines with a large degree of sweepback or even delta-shaped wings.

There are some designs in which the tail of the aircraft is located ahead of the wing. This type is termed tail-first or *canard* configuration.

Usually the elevator hinged at the rear of the tailplane is used to alter the tailplane lift. Moving the elevator up or down alters the camber of the tail surface and thus changes the lift.

In straight and level flight, the lift balances the weight and thrust is equal to drag.

The flight speed range of an aircraft can vary from the maximum speed that can be attained to the minimum speed at which the aircraft can be kept in the air, both without losing height.

The range of an aircraft is dictated by the engine, propeller, and the wind condition at the flying altitude.

The problem of converting the calorific value of the fuel by burning it with the oxygen in the air and then converting into mechanical work is governed by the law of thermodynamics. Even in the best engine, the efficiency of conversion of thermal energy to mechanical energy is just about 30%.

The lift-to-drag ratio will be a maximum when the drag is a minimum. The L/D ratio for the wing alone will be always greater than the L/D ratio of the entire aircraft.

The angle of attack that gives the best L/D ratio will be the same at all height or flying altitude and at all weight of the aircraft. Therefore, drag minimum simply implies the flight at the *best attitude* and has nothing to do with the density of air, or the payload.

If a flight is from place 1 to place 2 and back to place 2, the wind of any speed and any direction will adversely affect the range of action. The wind usually changes its direction and increases its velocity with height.

Endurance is the time duration for which an aircraft can stay in the air for a given quantity of fuel. For maximum endurance, the aircraft should fly at an angle of attack that gives the best value of $C_L^{3/2}/C_D$. The best value of $C_L^{3/2}/C_D$ for an aircraft will be at a greater angle of attack and a lower speed than for range.

Range of an aircraft is the distance it can cover in flight for a given quantity of fuel. The flying range of an aircraft will increase if it is flown faster than the minimum drag speed, but the drag will be slightly higher.

For jet engines, fuel consumption is approximately proportional to thrust. Therefore, for maximum endurance, it should fly with minimum thrust, that is, with minimum drag.

8

Gliding

8.1 Introduction

Gliding is a flight under the action of gravity and without the use of the engine. Thus of the forces, namely, the lift, L; weight, W; thrust, T; and drag, D, which usually act on an aircraft in flight, thrust is not acting on the machine in a gliding flight. Therefore, when an aircraft is in a steady gliding, it must be kept in a state of equilibrium by the lift, drag, and weight only. That is, in a gliding flight, the resultant of the lift and drag forces must be equal and opposite to the weight, as illustrated in Figure 8.1. As shown in the figure, the lift is at right angles to the glide path, and the drag is in the direction of the glide path but acts backwards. It is seen that

$$L = W \cos \alpha$$
$$D = W \sin \alpha$$

where α is the glide angle. Therefore,

$$\frac{L}{D} = \cot \alpha \tag{8.1}$$

This implies that the aerodynamic efficiency, L/D, during a glide will be high for small values of glide angle α. Thus:

- The aerodynamic efficiency of an aircraft during a glide depends on the glide angle. Therefore, smaller the glide angle, the more efficient the aircraft, and the farther the glide. However, we know that this is the same criterion as that of maximum range. Thus an aircraft that has a flat glide angle should also be efficient at flying for range, neglecting the influence of propulsion efficiency.
- For an aircraft to glide over a long distance, the angle of attack during the glide must be such that the L/D is a maximum. Usually, the angle of attack during a flat glide will be nearly the same as that during a straight and level flight and almost exactly the same as when flying for maximum range with piston engine.
- Gliding at an angle of attack greater or less than that gives the best L/D results in a steeper descent.

8.2 Angle of Glide

The gliding should be carried out *relative to the air*. Therefore, for an observer on the ground an aircraft gliding into the wind may appear to remain still or in some cases even to ascent. In such

Introduction to Aerospace Engineering: Basic Principles of Flight, First Edition. Ethirajan Rathakrishnan.
© 2021 John Wiley & Sons, Inc. Published 2021 by John Wiley & Sons, Inc.
Companion Website: www.wiley.com/go/Rathakrishnan/IntroductiontoAerospaceEngineering

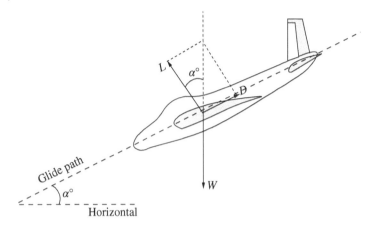

Figure 8.1 Forces acting on an aircraft during a glide.

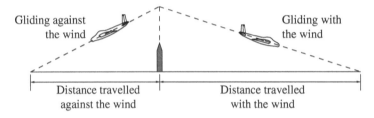

Figure 8.2 Wind effect on glide.

instances, the wind should have both a horizontal velocity and an upward velocity. For an observer on ground, an aircraft gliding against the wind will appear to glide more steeply relative to the ground. Also, when gliding with the wind, it will glide less steeply than the actual angle measured relative to the air. Therefore, an aircraft gliding with the wind will cover a larger distance and gliding against wind will cover lesser distance as illustrated in Figure 8.2.

8.3   Effect of Weight on Gliding

The gliding angle of an aircraft depends on the lift-to-drag ratio, which is independent of the weight. Therefore, it can be stated that the weight does not have any appreciable influence on the gliding angle, but the weight affects the air speed during the glide. In gliding without engine power, greater speed means higher drag, but now the thrust is provided by the component of the weight that acts along the gliding path, and thus the thrust will be higher for higher weight. Therefore, even though greater weight does not affect the glide angle, it does affect the speed.

8.4   Endurance of Glide

A glider, termed *sailplane*, must have a flat gliding angle if it has to cover a long distance. In other words, for covering any range from its starting point, a sailplane must have a *low rate of vertical descent* or *sinking speed* and must be able to stay long time in the air. The rate of vertical descent depends on both the angle of glide and air speed, during the glide. Therefore, to get a low rate of descent, the L/D should be large, that is, good aerodynamic design, and a low air speed, that is, low weight.

In actual gliding, lower rate of descent can be achieved by reducing the gliding speed below that which gives the flattest glide. Thus, there is an endurance speed for gliding just as for level flight.

8.5 Gliding Angle

The gliding angle in an important parameter that needs to be appropriate depending on the glide over the terrain. In other words, it should be noted that a flat glide angle is not always an advantage. For example, when approaching a small airfield near the edge of which there are high obstacles, it is necessary to reach the ground as soon as possible after passing over such obstacles. In these circumstances, a flat gliding angle is a disadvantage.

The gliding angle can be steepened by reducing the lift-to-drag, L/D, ratio. This can be done by increasing the angle of attack or by using an air brake, as illustrated in Figure 8.3. The air brake is a preferable means to increase L/D and may take the form of some kind of flap, but the modern tendency is to use various types of flap when lift is required and separate air brake or spoilers when drag is required.

8.6 Landing

Landing of an aircraft is a flight phase consisting of bringing the aircraft in contact with the ground at the lowest possible vertical velocity and, at the same time, somewhere near the lowest possible horizontal velocity relative to the ground.

It is essential that the velocity relative to the wind must be reasonably low. The first step in this direction is to land against the wind to reduce the ground run. However, the wind speed near the ground bound to be irregular due to environmental changes. If the wind speed suddenly decreases,

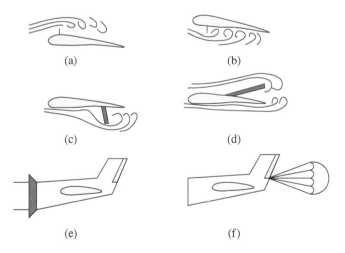

Figure 8.3   Some popular air brakes: (a) spoiler on wing top, (b) spoiler at the wing bottom, (c) split flap, (d) double flap, (e) spoiler around fuselage, and (f) tail parachute.

the aircraft due to its inertia will tend to continue at the same ground speed and will therefore lose air speed and, if already flying near the critical speed, may stall. Similarly, if the wind speed suddenly increases, the aircraft will temporarily gain air speed and will 'balloon' upwards, making it difficult to make contact with the ground at the right moment. Such instances may occur in changing and gusty winds, in up-currents caused by heating of parts of Earth's surface, in cases of turbulence caused by the wind flowing over obstructions such as hills, and due to the wind gradient. In a high wind of appreciable magnitude, and when flying a small light aircraft, these conditions may be dangerous, and the only solution at these situations is to land the aircraft at a higher speed than the usual landing speed.

The vertical velocity of landing can completely be reduced to zero, provided the forward velocity is sufficient for the wings to generate the required lift to balance the weight of the aircraft. The attitudes of an aircraft at various speeds and the corresponding angle of attack required to maintain level flight are illustrated in Figure 8.4.

Since the lift must be equal to weight, and $L = \frac{1}{2}\rho V^2 S C_L$, it is obvious that if V has to be as small as possible, the lift coefficient C_L must be as large as possible. The lift coefficient will be maximum near the stalling angle (about 15–20° in the case of an ordinary aerofoil); thus the maximum speed at which level flight can be maintained is just above the stalling speed.

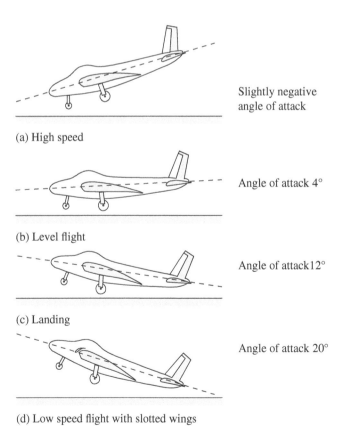

(a) High speed — Slightly negative angle of attack

(b) Level flight — Angle of attack 4°

(c) Landing — Angle of attack 12°

(d) Low speed flight with slotted wings — Angle of attack 20°

Figure 8.4 Attitudes of (a) maximum speed, (b) normal cruise flight, (c) normal landing, and (d) slow speed using flap and slots.

8.7 Stalling Speed

An aircraft may stall not only in level flight but also in gliding, climbing, or turning also. For example, when an aircraft is in a banked turn, the lift on the wings must be greater than the weight, and therefore the stalling speed is higher than the landing speed. Also, at any altitude, the air density will be less than the ground level density, and this means that to keep $\frac{1}{2}\rho V^2 S C_L$ equal to the weight, the stalling speed will be greater than that at ground level.

For airfields at high altitudes, such as in mountainous countries, the true landing speed of an aircraft will be considerably higher than at sea-level airfield. Also, in tropical countries, the air density is decreased, owing to the high temperatures, and the true landing speed is consequently increased. The take-off speed and the run required are also increased in both these instances.

When stalling intentionally, the aircraft is pulled into a steeply climbing attitude, and air speed allowed to drop to practically nil until the nose suddenly drops or one wing drops and the aircraft gets into dive or spin.

At this stage it is essential to note that it is not easy to define the stalling or stalling speed precisely. The stall occurs because the smooth airflow over the wing becomes separated. At small angles of attack, there is some turbulence near the trailing edge; as the angle increases, the turbulence spreads forward. The turbulence spreads in the span-wise direction also, usually from tip to root on highly tapered wings and from root to tip on rectangular wings. If the stall is defined as the breakup of the airflow, when would this occur? This can be answered as below.

There may be buffeting of the tailplane or main wing, but this too may be slight and unimportant or may be violent. As a result of the change from smooth to turbulent airflow, the curve of lift coefficient reaches a maximum and then begins to fall. We know that the stalling angle is the angle at which the lift coefficient is a maximum. Thus, the stall can be expected when the lift coefficient attains a maximum value, that is, when the aircraft flies near the minimum speed required for level flight. The minimum speed of most light aircraft is around 50 or 60 knots.

8.8 High-Lift Aerofoils

High-lift aerofoils are essentially aerofoils with some device by which the shape of the aerofoil can be altered during flight. Some of the commonly used high-lift devices are flaps and slots. The idea of varying camber with flaps and slots is old ones and is effective only for aircraft with low and moderate speeds. For high-speed aircraft, the camber has to be increased quickly. To meet this requirement, many devices have been developed. It is not easy to compare the movement of all these types of flaps and slots, because of the requirement of so many conflicting qualities they have to satisfy.

At low speeds, the device giving the highest maximum lift coefficient is the most suitable one. However, what is really required is a high-lift device that will be efficient in a low minimum speed, as well as a high maximum speed. This requirement implies that the device must be such that it can be altered from the position giving maximum lift (that is, flap down position) to the position of minimum drag (that is flap neutral position). Also, landing with low speed needs high lift, and the quick pull-up needs much drag, lift being of no consequence at these phases of operation. For a quick pull-up, an *air brake* that will assists the wheel brakes is necessary. However, an air brake cannot reduce actual landing speed; it can only improve the pull-up after landing. Once the aircraft is on the ground, the lift has to be quickly reduced to zero to achieve minimum wheel brake effectiveness; hence lift dumpers should be used after touchdown. The dumpers are devices that disrupt the flow over the top of the wing, resulting in the increase of drag and decrease of lift.

Another important problem to be addressed is the following. So far as landing is concerned, the attitude of the aircraft plays a dominant role. At this phase, the slots and flaps are at a disadvantage, since they attain their maximum lift coefficient at a greater angle of attack than the ordinary aerofoil. This means that to make full use of them, the angle of attack during the landing phase may need to be 25° or even more. Consequently, when an aircraft with a tailwheel type of undercarriage rests with its main and tail wheels on the ground, the angle of inclination of the wing is only about 15°. With a nosewheel type of undercarriage, the problem becomes worse. The following are the possibilities to land an aircraft at an angle of 25°.

- Allow the tailwheel to touch the ground before the main wheel, as shown in Figure 8.5a. However, this is hardly possible in reality.
- Have a high undercarriage, as shown in Figure 8.5b. This will cause extra drag and generally do more harm than good.
- Provide the main wing with a variable incidence gear similar to that which is sometimes used for tailplanes (Figure 8.5c). This involves a complex mechanical device.
- Set the wings at a much greater angle to the fuselage, as shown in Figure 8.5d. This means that in normal flight, the rear portion of the fuselage sticks up into the air at an angle that is inefficient from the drag point of view.

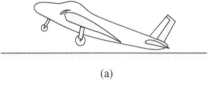

(a)

Figure 8.5 Different attitudes of aircraft during landing: (a) tail hitting first, (b) very high undercarriage, (c) variable incidence, (d) large riggers' angle of incidence.[1]

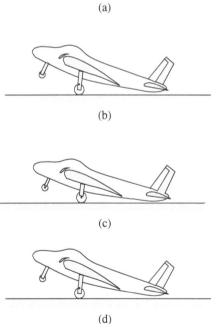

(b)

(c)

(d)

---

1 Riggers' angle is the angular distance between the chord and a line drawn longitudinally through the fuselage. It is determined during the design of the airplane and is the angle at which the wing is attached to the fuselage. Hence, it is a fixed angle.

8.9   Wing Loading

Wing loading of an aircraft is the ratio of the weight, W, to the wing planform area, S. In level flight, the weight is equal to the lift:

$$W = L$$

The lift is

$$L = \frac{1}{2}\rho V^2 S C_L$$

where ρ is the air density at the flight altitude, V is the flight speed, and C_L is the lift coefficient. Therefore,

$$W = \frac{1}{2}\rho V^2 S C_L$$

Thus the wing loading becomes

$$\boxed{\frac{W}{S} = C_L \frac{1}{2}\rho V^2} \tag{8.2}$$

From this expression for the wing loading, it is evident that for a given W, an increase in the wing area would reduce the wing loading and so also reduce the minimum velocity at which level flight is possible.

For changing the wing area, a device such as an actuator with gear arrangement is required. This will cause an increase in weight, W. Thus decreasing wing loading by increasing wing area, S, is practically not possible.

Equation (8.2) also shows that, other things being equal, the aircraft with low wing loading will have a lower minimum speed than one with a high wing loading. The aircraft with higher value of W/S will have higher landing speed. That is, it is not only the weight but also the ratio of weight to wing area that dictates the minimum speed. The wing loading of a sailplane may be less than $100\,\text{N/m}^2$, for a trainer aircraft the wing loading range may be from 300 to $1000\,\text{N/m}^2$, and for fighter, bomber, or commercial aircraft the wing loading may range from 1500 to $3000\,\text{N/m}^2$. In the modern design, the tendency is to increase wing loading by reducing wing area and then raising the maximum speed and then using flaps to keep down the landing speed.

8.9.1   Calculation of Minimum Landing Speed

By Eq. (8.2), we have

$$W = \frac{1}{2}\rho V^2 S C_L$$

For maximum lift coefficient, this becomes

$$W = C_{L_{max}} \frac{1}{2}\rho V^2 S$$

The designer has to decide on the important parameters such as the wing area, the type of aerofoil, and the landing speed at the early stages of the design itself. For deciding on the above parameters, the weight of the aircraft to be designed must be known. However, in the beginning for the design procedure, the weight needs to be guessed by the designer with his experience. From these discussions, it is evident that the landing speed is a vital parameter dictating the design of the aircraft right from the beginning.

Example 8.1

Find the landing speed of an aircraft with the mass of 1500 kg, flying at sea level, in knots, if the maximum lift coefficient of the wing is 1.2 and wing area is 60 m².

Solution

Given: $m = 1500$ kg, $C_{L_{max}} = 1.2$, and $S = 60$ m².

At sea level, the air density is $\rho = 1.225$ kg/m³.

By Eq. (8.2),

$$W = \frac{1}{2}\rho V^2 S C_L$$

During landing, $C_L = C_{L_{max}}$ and $V = V_{min}$. Thus,

$$W = C_{L_{max}}\frac{1}{2}\rho V^2_{min} S$$

Therefore,

$$V^2_{min} = \frac{W}{\frac{1}{2}\rho S C_{L_{max}}}$$

$$= \frac{mg}{\frac{1}{2}\rho S C_{L_{max}}}$$

$$= \frac{1500 \times 9.81}{\frac{1}{2} \times 1.225 \times 60 \times 1.2}$$

$$= 333.673$$

$$V_{min} = \sqrt{333.673}$$

$$= \boxed{18.27 \text{ m/s}}$$

That is, the landing speed is

$$V_{landing} = 18.27 \times 3.6$$

$$= \boxed{65.77 \text{ km/h}}$$

This speed can be expressed in terms of knots, noting that 1 km/h $\approx 1/(1.85)$ knot. Thus,

$$V_{landing} = 65.77/1.85$$

$$= \boxed{35.55 \text{ knot}}$$

8.10 Landing Speed

Landing speed of an aircraft is mainly dictated by the wing loading, W/S. Wing loadings are going up progressively. They went up slowly in the first three decades of flight and rather very slowly during the World War II. However, there is no sign of any halt in this process. Therefore, it is essential to presume that the wing loading will go up further.

With the invention of slots and flaps, the value of maximum C_L has gone up from just over 1 to about 3, or even 4, for modern aerofoil sections with slotted flaps and slots extending along 60% of the wing span. For $C_{L_{max}}$ of 1.22, a wing loading of 500 N/m² was considered high, but with a $C_{L_{max}}$

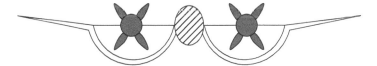

Figure 8.6 The concept of Custer Channel Wing.

of 3, even wing loading of 5000 N/m² has already been exceeded. A wing with $C_{L_{max}}$ of 1.22 and W/S of 500 N/m² gives a landing speed of 50 knots, whereas for $C_{L_{max}} = 3$ and $W/S = 5000$ N/m², the landing speed is about 100 knots.

At this stage, it is natural to think, can the wing loading be increased further? The increase in wing loading has a greater effect than the increase in $C_{L_{max}}$. Consequently, so far, this stringent requirement of increasing $C_{L_{max}}$ with W/S increase has been successfully addressed with improved flap design. However, it seems that there is not much scope to improve the flap any further. Therefore, to address the issue of increase in $C_{L_{max}}$, it is required to increase the wing loading further; therefore the designers are only left with the following options:

- Extend the flaps and slots along the whole span of the wing, perhaps also under the fuselage. This has already been done in some types of aircraft, where the ailerons also act as flaps, therefore known as '*flaperons*'.
- Dispense with the ailerons altogether and adopt an alternative form of lateral control, such as spoilers, or differential movement of the tail surface, thus known as '*tailerons*'.

These methods might give another 40% increase in $C_{L_{max}}$, and a landing speed will come down from about 100 to 85 knots for W/S of 500 N/m². At present with the aim of increasing $C_{L_{max}}$, many innovative concepts are proposed. Among them the good idea is that of using high-pressure air bled from the engine compressor to help induce attached flow over the flaps (blown flaps) at large deflections or even to produce a downward curtain of air that can turn the main airflow through extremely large angles: that is, the *jet flap*. Devices such as jet flaps are complex and heavy and therefore are used only for specialised aircraft, usually military aircraft. An example of a promising idea that may go to practical application is the Custer Channel Wing shown in Figure 8.6, in which the engines with pusher propellers will be suspended in the channels with the ailerons just on-board of the channels. The idea is simple; the propellers drew all over the curved wing section, thereby allowing the flow to remain attached at high angles at attack. However, there are two main problems associated with this idea. They are the following:

- The lifting surface also needs to have a good lift-to-drag ratio at high speed.
- As the minimum speed of the aircraft decreased, it becomes difficult to provide adequate stability and control by conventional means.

8.11 Short and Vertical Take-Off and Landing

In many situations low speeds or even vertical landing becomes mandatory. Until the twentieth century, the manual flights, such as balloons and airships, were made only in vertical take-off. However, there has been considerable research into both vertical and short take-off aircraft, leading to the extent of coining of two acronyms, VTOL for vertical take-off and landing and STOL for short take-off and landing. A hybrid of these two is termed STOVL, short take-off and vertical landing.

8.11.1 Gyroplane

The gyroplane is a flying machine capable of short take-off and landing, similar to a helicopter. The gyroplane is also referred to as *autogyro*. However, a gyroplane differs from a helicopter in the fact that in a helicopter the wings or blades are rotated by the power of the main engine, while in a gyroplane the rotating wings are driven only by the action of the air upon them and not by any power supply. Thus, forward speed is necessary in a gyroplane, as in a conventional aircraft. The forward speed for a gyroplane is provided by the thrust of an engine and propeller. An important feature of gyroplane rotor is its inclination backwards of the axis. The wings rotate automatically in such a way that even when the forward speed of the aircraft is far lower than the stalling speed of a conventional aircraft, the rotating wings are still striking the air at a considerable velocity and can thus provide sufficient lift to keep the aircraft in the air. In this way, the forward speed can be reduced to 5–10 knots that, in a slight head wind, means a ground speed is practically nil.

8.12 The Helicopter

Helicopter is essentially a rotary wing aircraft. In normal flight, the upward thrust of the revolving blades must be equal to the weight; forward motion is produced by inclining the effective axis of the rotor forward that normally entails the nose of the helicopter down. This is achieved by altering the cyclic variations in rotor blade incidence. The blade going into the wind gets more lift and drag than the blade going down the wind. This happens in any movement of the helicopter: forward, backward, or sideways. This becomes a serious problem when the helicopter moves at high speeds because the tip portions of the blade going into the wind meet compressibility problems before the helicopter itself is moving anywhere near the speed of sound. This effect is encountered for helicopter speeds near 370 km/h. Other problems associated with helicopter are the excessive vibration and noise compared with an aircraft of similar power.

Efforts to solve these problems resulted in the development of wing that not only will rotate but also have cyclic and collective pitch changing. Also, there are auxiliary propellers at the tail, or jets; some have jets at the wing tips too to rotate the blades. Though these devices help to solve the problem of vibration and noise to some extent, all of them add to the complications and weight of an already complicated helicopter machine.

The complexity associated with the flapping and drag hinges of the pitch changing mechanism has been replaced now by flexures. This technology reduced the weight of the rotor head. Another development is shrouding the tail rotor, leading to reduced drag and reducing the possibility of accidental contact with the tail rotor. Another recent advancement is replacement of the tail rotor with a variable air jet at the rear to perform the same function. This technique is termed *notar*, referring to no tail rotor. This is much quieter than the conventional helicopter and also avoids the possibility of damage to or from tail rotor.

However, the main limitation of low maximum speed of the conventional helicopter is not addressed. One of the reasons for this is that to generate lift, significant portions of the retreating blades must be going faster than the relative air speed past the helicopter. This means that portion of the advancing blade will be going more than twice as fast as the helicopter and will therefore go supersonic when the helicopter is still flying at speeds far below the speed of sound. Several solutions have been proposed to overcome this situation. One among them is the so-called *compound* helicopter. This has the novel features of a conventional helicopter but also has small wings and a separate engine or engines to provide forward speed directly. At high speed, the wings provide most of the lift, allowing the rotor to be rotated relatively slowly or even feathered.

8.13   Jet Lift

Jet lift and *thrust vectoring* aircraft is an alternate to the helicopter. The Harrier, known as the AV-8 in the United States, is a successful jet lift aircraft. The harrier uses downward deflection of the engine to produce a lifting thrust until sufficient forward speed has been obtained for the wings to take over the job of lifting. The engine outlet nozzles, two from the hot exhaust and two from the compressor, can be *vectored* (varied in outlet angle) from vertical or even slightly forward to horizontal to obtain the desired balance between lift and forward thrust. The critical problem of providing stability and control at low speed is solved by using *puffer* jets of compressed air bled from the engine. These are located at the nose, the tail, and both wing tips. Like the helicopter, this aircraft can be flown backwards and sideways. It can also hover and land and take off vertically. Although capable of vertical take-off, it is more economical in terms of fuel burnt for the aircraft to make a short horizontal take-off, with the engines set at an intermediate angle so as to provide thrust and some lift.

8.14   Hovercraft

Hovercraft is a vehicle that lifts above the ground or water surface by means of a fan, which produces a raised pressure under the vehicle. Leakage of the air from the underside is restricted by a peripheral 'curtain' to high-speed air, normally supplemented by a flexible 'skirt'.

Another type of vehicle that flies just above the surface is the skimmer or ekranoplan. These vehicles resemble a conventional aircraft but by flying just above the surface take advantage of the 'ground effect' that greatly reduces the wing downwash and consequently reduces the accompanying trailing vortex drag. The main disadvantages of the ekranoplan are the problems of ensuring the correct ground clearance at all times and the very large amount of engine power needed to get the aircraft to lift off from water.

8.15   Landing

Landing is a flight phase involving descend from cruise altitude, gliding in a straight line, flattening out to a straight line parallel to ground, and gentle touchdown, as illustrated in Figure 8.7. During landing essentially the wing loading increases. With increase in wing loading, the landing speed has to become higher, and this demands that the flaps need be used to meet the increased lift requirement.

While approaching for landing, the aircraft may undershoot or overshoot. If the aircraft undershoots in spite of the best gliding angle, the pilot has nothing to do but land. This landing with undershooting is considered bad flying, if the engine was functioning satisfactorily. In the anxiety to avoid undershooting, there is a tendency to overshoot, especially since it would seem to be easy to loose any unnecessary height. In the case of overshooting, to loose height at the required rate and overshoot, usually the following methods are available:

- Sideslipping.
- Prolonging the glide by S-turn.
- Putting the nose down and gliding fast.

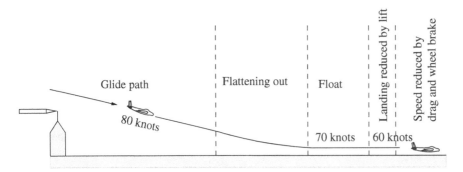

Figure 8.7   Illustration of landing phases of flight.

- Holding the nose up and gliding slowly.
  The landing process of a typical aircraft consists of five phases, namely, the glide, the flattening out, the float or hold-off, the landing, and the pull-up, as illustrated in Figure 8.7. A closer look at these phases of landing will reveal that in all of these, the flaps have their part to play.
- It should be noted that the glide during the last phase of about 150 m should be straight, without any slipping to turning to one side or the other. This is possible only by means of controlling the gliding angle relative to the earth without unduly raising or lowering the air speed. Flaps can be used to do this, at any rate over a limited range of gliding angle. As the flaps are lowered, both lift and drag increase. The increase in lift tends to flatten the gliding angle and make the aircraft to glide at a slower air speed without approaching closer to the stalling speed. The increase in drag tends to steepen the glide angle, and gliding attitude, of the aircraft for some air speed. The net effect depends on whether the L/D ratio is raised or lowered. In other words, it can be stated that the flaps give some *control over the gliding angle.*
- The *flattening out* flight mode involves a change of direction, an acceleration, and force towards the centre of the curved path. This force must be provided by the wings. Therefore, the wings should have more speed and more angle of attack. These increased speed and angle lead to higher stalling speed. The steeper the original glide, the greater the change in flight path involved; the aircraft should have more speed for flattening out. This means that the steeper the glide, the faster must be the glide speed. However, faster gliding speed is not desirable.
- After flattening out, the excess speed should be reduced – this may be called *float* or *hold-off*. In this flight mode, the drag of flaps or air brakes play their part, as do the wings themselves as they are brought to the angle for actual landing.
- The last phase in the landing approach is termed *landing.* The landing phase is momentary only, but the landing speed is of the utmost importance because it settles both the gliding speed and the distance to pull up after landing. At this stage it is essential to note that the drag, whether caused by flaps or anything else, cannot reduce landing speed.
- After the landing comes the *pull-up phase.* For a quick pull-up, drag is the essential component required. The drag provided by wheel *brakes* and air *brakes* is the better suited drag, provided the aircraft can stand it and does not tip on its nose. In addition to air brakes, some type of flap, when fully lowered, gives good braking effect, and so do the wings at their angle of 16° or so with a tailwheel type of undercarriage.

For aircraft with nosewheel type undercarriage, there is lack of air drag during landing. An effective substitute for this disadvantage is a tail parachute. The air-braking effect is greatest at the beginning of the landing run; later on the wheel braking can be used to safely bring the aircraft to rest.

Brake application is a straightforward mechanism, but apart from the question of coefficient of friction between wheels and ground and the serious danger of 'aquaplaning'.[2] When there is water on the runway, there are some aspects of brake that are peculiar to aircraft. For propeller aircraft, the centre of gravity is high above the wheels; due to this, for tailwheel type, there would be a tendency to go over on to the nose if the brakes are applied suddenly. Furthermore, if a tailwheel aircraft starts to swing, the centre of gravity behind the wheels will cause the turning to increase. This may be checked by the differential action of the brakes. However, the tricycle or nosewheel undercarriage can remove the cause for this trouble. For nosewheel undercarriage, the centre of gravity is in front of the main wheels, and there is to tendency to swing.

8.16   Effect of Flaps on Trim

Raising or lowering of flaps affects the airflow over both upper and lower surfaces of the wing and ahead of and behind the wing too, as illustrated in Figure 8.8.

The airflow in turn affects the pressure distribution and force and moments on the wing and the tailplane. When the flap is lowered, the air flows faster over the upper surface of the wing. This would cause higher suction and make the centre of pressure on the top surface to move forward. This would cause tail-heaviness.

The downwash caused by flap down will be large. If the tail is so situated as to receive the full advantage of this downwash, there will be downward force on the tailplane, leading to tail-heaviness.

For a low-wing aircraft, the low position of the drag acting on the flap, especially when fully lowered, will tend towards nose-heaviness. For a high-wing aircraft, the high portion of the drag may tend towards tail-heaviness.

The flap effect on the pitching moment depends entirely on the type of flap or slots used, on how much they are lowered, and on the situation of the tailplane. Slotted flaps, and flaps that move backwards resulting in the increase of the rear portion of the wing area, will nearly always cause a nose-down moment that sometimes has to be counteracted by leading edge slots and flaps.

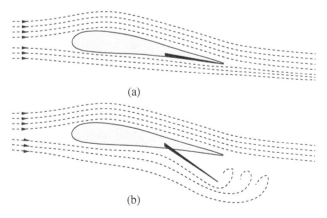

(a)

(b)

Figure 8.8   Flow around a wing with flap (a) neutral and (b) deflected down.

---

2 Aquaplaning, also known as hydroplaning, is a condition in which standing water, slush or snow, causes the moving wheel of an aircraft to lose contact with the load-bearing surface on which it is rolling with the result that braking action on the wheel is not effective in reducing the ground speed of the aircraft.

Sometimes, the change of trim is in one direction for the first part of the lowering of the flap, usually tail-heavy, and in the other direction when full flap is lowered, usually nose-heavy. In some aircraft, the effects cancel each other that there is little or no change of trim.

It is essential to note that the technique of landing a modern high-performance aircraft is totally different from the landing of a light aircraft. For modern aircraft, the speed, height, angle of descent, and a set of other factors such as the flap and power setting must be correct within very close limits as the aircraft crosses the airfield threshold. Manual landing of a modern aircraft requires a great deal of skill, and the majority of landing nowadays are made by automatic control.

8.17 Summary

Gliding is a flight under the action of gravity and without the use of the engine. When an aircraft is in a steady Gliding, it must be kept in a state of equilibrium by the lift, drag, and weight only.

This implies that the aerodynamic efficiency, L/D, during a glide will be high for small values of glide angle α.

The gliding should be carried out *relative to the air*.

The gliding angle of an aircraft depends on the lift-to-drag ratio, which is independent of the weight.

For covering any range from its starting point, a sailplane must have a *low rate of vertical descent* or *sinking speed* and must be able to stay a long time in the air.

The gliding angle can be steepened by reducing the lift-to-drag, L/D, ratio.

Landing of an aircraft is a flight phase consisting of bringing the aircraft in contact with the ground at the lowest possible vertical velocity and, at the same time, somewhere near the lowest possible horizontal velocity relative to the ground.

An aircraft may stall not only in level flight but also in gliding, climbing or turning also. For example, when an aircraft is in a banked turn, the lift on the wings must be greater than the weight, and therefore the stalling speed is higher than the landing speed. Also at an altitude, the air density will be less than the ground level density, and this means that to keep $\frac{1}{2}\rho V^2 S C_L$ equal to the weight, the stalling speed will be greater than that at ground level.

For airfields at high altitudes, the true landing speed of an aircraft will be considerably higher than an sea-level airfield. Also, in tropical countries the air density is decreased, owing to the high temperatures, and the true landing speed is consequently increased.

High-lift aerofoils are essentially aerofoils with some device by which the shape of the aerofoil can be altered during flight. Some of the commonly used high-lift devices are flaps and slots.

Wing loading of an aircraft is the ratio of the weight, W, to the wing planform area, S.

Landing speed of an aircraft is mainly dictated by the wing loading, W/S. For $C_{L_{max}}$ of 1.22, a wing loading of 500 N/m^2 was considered high, but with a $C_{L_{max}}$ of 3, even wing loading of 5000 N/m^2 has already been exceeded. A wing with $C_{L_{max}}$ of 1.22 and W/S of 500 N/m^2 gives a landing speed of 50 knots, whereas for $C_{L_{max}} = 3$ and $W/S = 5000$ N/m^2, the landing speed is about 100 knots.

The gyroplane is a flying machine capable of short take-off and landing, similar to a helicopter. A gyroplane differs from a helicopter in the fact that in a helicopter the wings or blades are rotated by the power of the main engine, while in a gyroplane the rotating wings are driven only by the action of the air upon them and not by any power supply. Thus forward speed is necessary in a gyroplane, as in a conventional aircraft.

Helicopter is essentially a rotary wing aircraft. In normal flight, the upward thrust of the revolving blades must be equal to the weight; forward motion is produced by inclining the effective axis of the rotor forward that normally entails the nose of the helicopter down. This is achieved by altering the cyclic variations in rotor blade incidence.

Hovercraft is a vehicle that lifts above the ground or water surface by means of a fan, which produce a raised pressure under the vehicle.

Another type of vehicle that flies just above the surface is the skimmer or ekranoplan. These vehicles resemble a conventional aircraft but by flying just above the surface take advantage of the 'ground effect' that greatly reduces the wing downwash and consequently reduces the accompanying trailing vortex drag.

Landing is a flight phase involving descend from cruise altitude, gliding in a straight line, flattening out to a straight line parallel to ground, and gentle touchdown.

9

Performance

9.1 Introduction

Performance of an aircraft consists of take-off, climb, cruise or level flight, descend or gliding, and landing. Level flight is the standard condition of flight with which all other manoeuvres are compared. Gliding involves simple fundamental principles that are more elementary than those of level flight. Landing is something special involving principles of flight at low speeds.

9.2 Take-Off

One of the main problems associated with take-off is to keep the aircraft on a straight and narrow path. This difficulty applies mainly to propeller-driven aircraft. The main requirement during take-off is to obtain sufficient lift to support the weight with the least possible run along the ground. To meet this lift requirement, the angle of attack is kept small during the first part of the run so as to keep the drag small. Then, when the speed has reached the minimum speed of flight, if the tail is lowered and the wings brought to about 15° angle of attack, the aircraft probably leaves the ground with the least possible run. However, this mode of flight with 15° angle of attack is dangerous because, once having left the ground, any attempt to climb by further increase of angle will result in stalling and drop back to the ground. Therefore, it is necessary to allow the speed to increase beyond the stalling speed before 'pulling off', and sometimes the aircraft is allowed to continue to run in the tail-up position until it takes off of its own accord.

The take-off process is strongly influenced by the runway surface. To reduce the length of run, an increase in the angle of climb after leaving the ground – so as to clear obstacles in the outskirts of the airfield – the take-off will, when possible, be made against the wind. Other aids to take-off are slots, flaps, and any other devices that increase the lift without unduly increasing the drag.

During take-off, whether the flap can be used or not depends on whether the increase of lift caused by flap deflection, leading to the decrease of take-off speed, makes up for the reduction in the acceleration caused by the increased drag associated with the flap deflection. The lift increase gained with flap deflection helps the take-off considerably if the deflection is nearly always used for take-off speed otherwise required.

Other problems associated with the take-off are the following. The undercarriage may take away nearly during flight, but when lowered they are less streamlined than the fixed undercarriage, and their drag may hamper the take-off quite considerably. The lower undercarriage that can be used with jets is a great advantage in this respect. With high wing loading, the take-off speeds go up, and the length of run needed to attain such speeds is liable to become excessive.

Introduction to Aerospace Engineering: Basic Principles of Flight, First Edition. Ethirajan Rathakrishnan.
© 2021 John Wiley & Sons, Inc. Published 2021 by John Wiley & Sons, Inc.
Companion Website: www.wiley.com/go/Rathakrishnan/IntroductiontoAerospaceEngineering

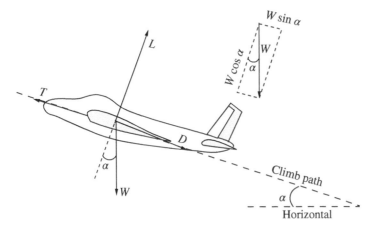

Figure 9.1 Forces acting on an aircraft during a climb.

9.3 Climbing

During level flight the power of the engine must produce a thrust equal to the drag of the aircraft at that flight speed. At this mode if the engine has some reserve power and if the throttle is further opened, either:

(i) The pilot can put the nose down slightly and maintain level flight at an increased speed and decreased angle of attack, or
(ii) The aircraft will commence to climb.

If the path travelled by the aircraft is in the same direction as the thrust, the forces acting on the aircraft will be as shown in Figure 9.1.

The forces acting on the aircraft in the direction parallel and normal to the direction of flight, respectively, are

$$T = D + W \sin \alpha \qquad (9.1)$$

$$L = W \cos \alpha \qquad (9.2)$$

where T is the thrust, D is the drag, W is the weight, and α is the angle of climb.

From Eq. (9.1), it is seen that, during a climb, the thrust required is greater than the drag. Also, the thrust required increases as the climbing angle increases. For $\alpha = 0$, Eqs (9.1) and (9.2) simplify to

$$T = D$$

$$L = W$$

These are forces acting on the aircraft in the direction of flight and normal to the flight, during a level flight.

9.4 Power Curves: Propeller Engine

Performance curves are plots of engine power versus aircraft flight speed. The climb performance of an aircraft can be analysed using the performance curves. Many interesting features of the performance can be studied by estimating the power available from the engine and the power required for the level flight at various speeds.

The procedure for the analysis of the performance of jet and rocket systems of propulsion is different from the present analysis of propeller engine performance because for jet and rocket engines the thrust is of primary importance than the power. Therefore, the propulsion of jet and rocket engines should be dealt with separately. In this section, our focus is essentially on the performance of piston-engine aircraft.

The engine power required during flight is given by the product of the total drag, D, and the flight speed, V:

$$\text{Power} = D \times V \tag{9.3}$$

The total drag is the sum of the drag of the wing (also called the profile drag) and the drag due to the parts of the aircraft other than the wing (usually referred to as the parasite drag). Theoretically the profile and parasite drags can be computed separately and added to get the total drag. Another way of finding the total drag is by measuring the drag of the model of the complete aircraft in a wind tunnel and scaling up to full size.

The power available is the power given by the engine to the propeller termed *brake power* multiplied by the efficiency of the propeller, η_p. Thus,

$$\text{Power available} = \text{Brake power} \times \eta_p \tag{9.4}$$

The propeller efficiency is usually of the order of 80%. A typical plot of power available to power required is shown in Figure 9.2.

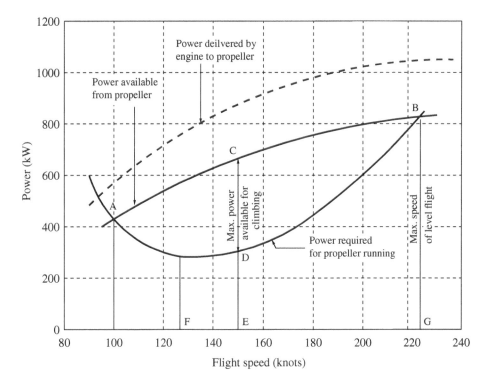

Figure 9.2 Power available and power required curves as a function of flight speed.

9.5   Maximum and Minimum Speeds in Horizontal Flight

From the plot of power available and power required curves shown in Figure 9.2, many interesting results can be inferred. In the speed range from point A to B, in Figure 9.2, where the power available is more than the power required, level flight is possible. The speed at points A and B, respectively, corresponds to the minimum and maximum speeds at which level flight is possible. Between points A and B, the difference between the power available and power required at any particular speed represents the amount of extra power that can be used for climbing at that speed. Thus, the best rate of climb for the aircraft whose power available and power required curves are those shown in Figure 9.2 is at flight speed of about 75 m/s.

Another feature that can be inferred from Figure 9.2 is that the speed corresponding to the minimum speed required, point C, is the speed for maximum endurance.

9.6   Effect of Engine Power Variation

For a fixed forward speed of the aircraft, the power available is a fixed quantity. However, the power of the engine can be varied considerably by manipulating the engine controls. If the power available curve shown in Figure 9.2 represents the power at some economical condition of engine operation with lean (weak) mixture, then more power can be obtained using rich mixture, and the absolute maximum power by operating the throttle to the maximum possible boost and using the maximum possible rpm – with fixed-pitch propellers this is the case of full throttle.

Only at special circumstances the full power of the engine should be used, that too only for a limited time. The engine will be damaged if it runs at full power for a long time. From the point of view of the aircraft, it makes no difference whether the power is decreased by reducing boost, or lowering the rpm, or both, but for fuel economy it is generally advisable to lower the rpm. As the power is reduced, the minimum speed of level flight becomes slightly greater, the maximum speed becomes considerably less, and the possible rate of climb decreases at all speeds.

Another point to be noticed is that the lowest speeds can be obtained with the engine running at full throttle. However, this flight condition cannot easily be sustained in practice because a small decrease in speed would mean an increase in required power and a simultaneous decrease in available power.

As the engine is throttled down, a state is reached at which there is only one possible speed of flight. This is the speed at which the least engine power will be used and at which the maximum endurance can be achieved.

9.7   Flight Altitude Effect on Engine Power

The flight performance of an aircraft is strongly influenced by the altitude at which it flies. Therefore, it is natural to ask whether it is preferable to fly high or fly low when travelling from one place to another. For this question there is no direct answer because there are many conflicting considerations, which have to be taken into account. Some of them are the temperature, wind speed, and quantity of oxygen in the atmospheric air. In addition to these, the performance of the aircraft must also be considered to decide on the suitable flight altitude.

Now let us focus on how the performance will be affected as the altitude of flight is increased. We know that as the altitude increases the density of air decreases. Therefore, to support the same aircraft weight while maintaining the same lift coefficient and altitude, it is necessary to increase the speed so as to maintain the same dynamic pressure. This means that we need to fly at the *same indicated air speed*. This also implies that the drag will remain the same as before. Thus, from the point of view of the airframe, it appears at first as though there is no disadvantage in flying high. Indeed the aircraft can actually fly faster for the same amount of drag and thrust. The main issue associated with flying high comes from the engine and propulsion system, listed below:

- The power required is the product of the drag and the speed. Therefore, although the drag and thrust can remain the same when the height is increased, the increase in the speed implies that the power required will increase.
- The power output of a piston engine falls as the air density decreases, and although this can be compensated by adding a supercharger or turbocharger, there is a limit to the amount of super-charging that is practical to use.
- Since higher speed is required to maintain the same lift and drag, the propeller will also have to go faster and will run into the problem of compressibility effects.
  In addition to these propulsion problems, the following problems of airframe and operational aspects also must be addressed:
- The cockpit and cabin need to be pressurised, which adds to the complexity and weight.
- Increase of height and speed will result in compressibility problems associated with the flow past the aircraft and the propeller.

Whatever be the efforts made to overcome the difficulties associated with flight at high altitudes, in propeller-driven aircraft, the general tendency for the power available to decrease and power required to increase with altitude remains. A typical plot illustrating the effect of altitude on the power available and power required is shown in Figure 9.3.

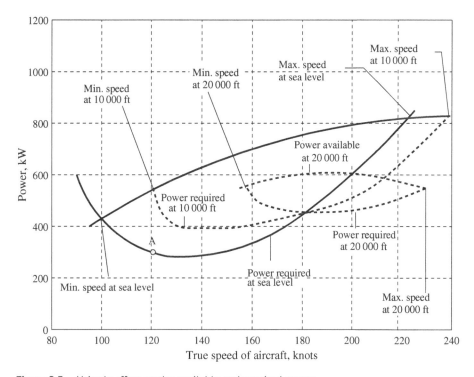

Figure 9.3 Altitude effect on the available and required power.

The effect of decrease in power available and increase in power required with altitude will cause the curves to close in towards each other, resulting in a gradual *increase in the minimum speed and a decrease in the maximum speed.*

It is essential to note that the air speed indicator is also affected by the change in density and consequently reads lower than the true air speed. This is what accounts for the power required curve moving over to the right as the altitude increases. That is, the power required curves for 4000 and 8000 m would simply displaced upwards compared with that at sea level. The difference between true and indicated speed also accounts for another discrepancy in that; the curves as plotted against true air speed suggest that the air speed to give the best rate of climb *increases* with height. However, the indicated speed for the best rate of climb *falls* with height.

9.8 Ceiling

Ceiling is the flying altitude at which there is only one possible speed for level flight and the *rate of climb is zero*. At the ceiling altitude, the engine does not have any extra power, that is, the power available is fully used. Thus, the ceiling is of little use for practical purposes, and therefore the idea of *service ceiling* is introduced. The service ceiling is that height at which the rate of climb becomes less than 0.5 m/s or some other specified rate.

9.9 Effect of Weight on Performance

When the weight is increased, the lift will also have to be increased. Therefore, either the aircraft must fly at a larger angle of attack, or the angle of attack is kept the same, at a higher speed. This speed can be calculated as follows.

Let the old weight be W_1 and the new weight be W_2. Let the old speed be V_1 and the new speed be V_2. Since the angle of attack is the same, the lift coefficient, C_L, will be the same at both speeds. Therefore,

$$W_1 = \frac{1}{2}\rho V_1^2 S C_L$$
$$W_2 = \frac{1}{2}\rho V_2^2 S C_L$$

where ρ is the density of air and S is the wing area. From these relations for W_1 and W_2, we get

$$\frac{V_2}{V_1} = \sqrt{\frac{W_2}{W_1}} \tag{9.5}$$

The effects due to the additional weight on the aircraft performance are the following:

- Slight reduction in maximum speed.
- Large reduction in rate of climb.
- Increase in minimum speed.

Note that these effects are the same as the effect of an increase of altitude. However, in spite of the similarity in effect of increase of weight and increase of altitude, the increase in weight does not affect the reading of the air speed indicator.

Because of the problems associated with operating piston engines at high altitudes, they are usually employed for aircraft designed for flight at relatively low altitudes. Thus the piston engine is used only for light general aviation. For propeller propulsion at high altitudes and high speeds, gas-turbine-based *turboprop propulsion* is more appropriate because the efficiency at high speed has been addressed with some success in recent years. With very advanced propellers, it is possible to fly even at transonic speeds. High-speed propellers have swept tips. The efficiency of propeller propulsion is theoretically greater than that of a pulse jet. Because of this, the turboprop is widely used for small regional airliners, transport aircraft, and military trainers, where flight at very high altitude and speed are not necessary. In addition to the advantages in terms of fuel consumption, the use of propellers does not add to the weight, cost, complexity, and maintenance requirement.

Example 9.1

If an aircraft weighing 62 kN has to carry an extra load of 15 kN, keeping the angle of attack the same, what should be the extra power required?

Solution

Given: $W_1 = 62$ kN and $W_2 = 62 + 15 = 77$ kN.

By Eq. (9.5),

$$V_2/V_1 = \sqrt{W_2/W_1}$$
$$= \sqrt{77/62}$$
$$= 1.114$$

Since the angle of attack is the same, the lift-to-drag ratio remains constant. That is,

$$L_1/D_1 = L_2/D_2$$

Thus,

$$D_2 = (L_2/L_1)\, D_1$$
$$= (77/62)D_1$$
$$= 1.242\, D_1$$

The corresponding power P_2 is

$$P_2 = D_2 V_2$$
$$= (1.242\, D_1) \times (1.114\, V_1)$$
$$= (1.242 \times 1.114)\, D_1 V_1$$
$$= 1.384\, P_1$$

Thus the extra power becomes

$$\frac{P_2 - P_1}{P_1} = \frac{P_2}{P_1} - 1$$
$$= 1.384 - 1$$
$$= 0.384$$
$$= \boxed{38.4\ \%}$$

9.10 Jet Propulsion Effect on Performance

For a jet engine, thrust and fuel consumption do not change much with speed. Therefore, the speeds for optimum range and endurance at any fixed altitude are faster than the speeds for a propeller-driven aircraft with the same airframe. Also, by flying higher, the aircraft will fly faster for a given drag and thrust. This implies that the *optimum* speeds for range and endurance will increase with altitude. Luckily, the efficiency of the jet also increases with altitude. Thus with turbojet propulsion, not only the optimum speeds increase with height, but also the distance that the aircraft can fly for each kilogram of fuel increases. Thus, for a jet aircraft, it is advantageous in terms of range and fuel efficiency to fly high and fast. The limitation of maximum speed is due to the onset of severe compressibility effects, as the speed nears transonic range. The feature of best economy at high speed is an important reason for the popularity of jet propulsion for airliners. Other advantages are the low noise and only a marginal vibration in the cabin.

Some of the important differences between the jet and piston-engined aircraft are the following:

- On a piston-engined aircraft, the engine power is regulated by the throttle control, because throttling controls the airflow in to the engine.
- On a jet aircraft, the engine is regulated by adjusting the fuel flow, and it is the thrust that is controlled directly, rather than the power.

Another difference between the piston and jet aircraft is that the higher speed of the jet aircraft means that the speed relative to the speed of sound becomes very important. Therefore, for jet engine performance estimates, the data needed to be displayed in a different form than that used for piston aircraft. It is better to work in terms of *thrust* and *drag* rather than power required and power available, and we need to know how the thrust and drag vary with *Mach number* and altitude. That is, the simple performance calculations that are used for low-speed piston aircraft are not appropriate for high-speed jet aircraft.

9.11 Summary

Performance of an aircraft consists of take-off, climb, cruise or level flight, descend or gliding, and landing.

The take-off process is strongly influenced by the runway surface. During take-off whether the flap can be used or not depends on whether the increase of lift caused by flap deflection, leading to the decrease of take-off speed, makes up for the reduction in the acceleration caused by the increased drag associated with of the flap deflection.

During level flight, the power of the engine must produce a thrust equal to the drag of the aircraft at that flight speed.

Performance curves are plots of engine power versus aircraft flight speed. The climb performance of an aircraft can be analysed using the performance curves.

The engine power required during flight is given by the product of the total drag, D, and the flight speed, V:

$$\text{Power} = D \times V$$

The power available is the power given by the engine to the propeller termed *brake power* multiplied by the efficiency of the propeller, η_p. Thus,

$$\text{Power available} = \text{Brake power} \times \eta_p$$

The flight performance of an aircraft is strongly influenced by the altitude at which it flies.

Ceiling is the flying altitude at which there is only one possible speed for level flight and the *rate of climb is zero*.

For a jet engine, thrust and fuel consumption do not change much with speed. Therefore, the speeds for optimum range and endurance at any fixed altitude are faster than the speeds for a propeller-driven aircraft with the same airframe.

On a piston-engined aircraft, the engine power is regulated by the throttle control, because throttling controls the airflow in to the engine. On a jet aircraft, the engine is regulated by adjusting the fuel flow, and it is the thrust that is controlled directly, rather than the power.

10

Stability and Control

10.1 Introduction

The ability of the aircraft to return to the same flying mode, when slightly disturbed from that condition, without any effort on the part of the pilot is called *stability*. An aircraft may be stable under some flight conditions or unstable under other conditions. For example, an aircraft that is stable during straight and level flight may be unstable when inverted. If an aircraft were stable during nosedive, it would mean that it would resist effects on the part of the pilot to extricate it from nosedive. The stability is also referred to as *inherent stability*. Some military combat aircraft are intentionally made to be inherently unstable to increase their manoeuvrability. This calls for sophisticated automatic stabilisation system, which is reliable. For civil aircraft, inherent stability is not permitted.

Stability is different from balance or *trim* of an aircraft. An aircraft that flies with one wing lower than the other may often, when disturbed from this attitude, return to it. Such an aircraft is out of its proper trim, but it is not unstable.

An aircraft that tends to return to its original position, when disturbed, is said to be stable. An aircraft that tends to move further away from the original position, when disturbed, is said to be *unstable*. However, if the disturbance sets the aircraft in a new position, the aircraft is said to have *neutral stability*.

Some of the ways in which an aircraft left to itself will behave are illustrated in Figure 10.1, showing only the pitching motion. The same considerations apply to roll and yaw, although an aircraft may have quite different stability characteristics about its three axes. Figure 10.1a shows complete dead-beat stability, which is very rarely achieved in practice. An unusual type of stability and oscillation that is gradually damped out is illustrated in Figure 10.1b. The steady oscillation shown in Figure 10.1c is a form of neutral stability. An instability in which the oscillation steadily grows worse is illustrated in Figure 10.1d.

The degree of stability would differ according to the *stick-fixed* and *stick-free* conditions. For example, in pitching stick-fixed means that the elevators are held in their neutral position relative to the tailplane, whereas stick-free means the control column is released allowing the elevators to take up their own positions.

Another factor that influences the stability is the *power-off* or *power-on* condition. On modern aircraft, the engine thrust is of the order of or even greater than the airframe weight and therefore may significantly influence the stability.

The word 'control' means the power of the pilot to manoeuvre the aircraft into any desired position.

Introduction to Aerospace Engineering: Basic Principles of Flight, First Edition. Ethirajan Rathakrishnan.
© 2021 John Wiley & Sons, Inc. Published 2021 by John Wiley & Sons, Inc.
Companion Website: www.wiley.com/go/Rathakrishnan/IntroductiontoAerospaceEngineering

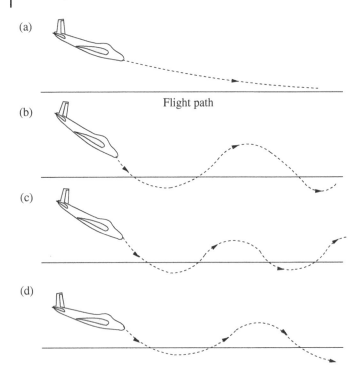

Figure 10.1   Illustration of flight modes an aircraft left to itself would experience.

The stability or control of an aircraft concerning *pitching* about the lateral axis is called *longitudinal stability* or *control*, respectively. Stability or control concerning rolling about longitudinal axis is called *lateral stability* or *control*, respectively. Finally, stability or control concerning yawing about the normal axis is called *directional stability* or *control*.

10.2   Longitudinal Stability

To obtain stability in pitching, it is essential to ensure that if the angle of attack is increased, forces will act in such a way as to depress the nose, and this decreases the angle of attack. An unswept wing with a cambered aerofoil section cannot be stable or 'trimmed' to give positive lift or at the same time be stable in the sense that a positive increase in incidence produces a nose-down pitching moment about the centre of gravity.

The position of centre of gravity with respect to the wing itself can be improved to some extent by *sweepback*, by *washout* towards the wing tips, by change in wing section towards the tips, and by a reflex curvature towards the trailing edge of the wing section.

It is not only the wing that affects the longitudinal stability of the aircraft as a whole, but also in general it can be said that this is dependent on the following:

- The position of the centre of gravity, which must not be too far back. This is probably the most important consideration.

10.3 Longitudinal Dihedral

The angle between the tailplane chord and wing chord is called *longitudinal dihedral*, as illustrated in Figure 10.2. The longitudinal dihedral is regarded as an important characteristic dictating the longitudinal stability of most type of aircraft. However, it cannot be said that an aircraft that does not possess this feature is necessarily unstable longitudinally, because so many considerations other than longitudinal dihedral enter into the problem of longitudinal stability. In any case, it is the actual angle at which the tailplane strikes the airflow that matters; therefore, the downwash from the wings also play a dominant role on the longitudinal stability. This downwash, if the tailplane is in the stream, will cause the angle of attack to be less than the angle at which the tailplane is set. For this reason, even if the tailplane is set at the same angle as the wings, this will *in effect* be a longitudinal dihedral angle, and this may help the aircraft to be longitudinally stable.

Let us consider an aircraft flying with angle of attack of the wing as 4° and the angle of attack of the tailplane as 2°. Let a sudden gust cause the nose to rise, inclining the longitudinal axis of the aircraft by 1°. Let us analyse the effect of this gust on the aircraft's stability. The momentum of the aircraft will cause it temporarily to continue moving practically in its original direction and at its previous speed. Therefore, the angle of attack of the wing will become nearly 5° and the tailplane nearly 3°. The pitching moment about the centre of gravity of the wing will probably have a nose-up, that is, unstable tendency; the pitching moment of the tailplane will have a nose-down tendency. If the restoring moment caused by the tailplane is greater than the upsetting moment caused by the wings, thus the aircraft will become stable.

When the tailplane is in front of the wings, the tail must have greater angle than the wings to have longitudinal dihedral. The wing will still be at an efficient angle, such as 4°, so that the tail surface may be at, say, 6° or 8°. Thus, the tail is working at a very inefficient angle and will stall at some few degrees sooner than the wings. This may be regarded as an advantage, since the stalling of the front surface (tail surface) will prevent the nose being raised any further, and therefore the wing will never reach the stalling angle.

For a tail-less aircraft, there is no separate surface either in front or behind, the wings must be heavily swept back, and there is a 'washout' to decrease the angle of incidence as the wing tip is approached, so that these wings tips are in effect, as in exactly the way as the ordinary tailplane.

Figure 10.2 Longitudinal dihedral angle.

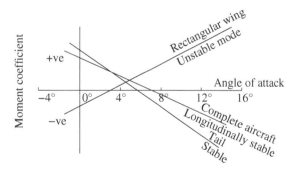

10.4   Lateral Stability

For lateral stability, the aircraft must have the capability to restore its original mode of flight, when it is subjected to a slight roll. For an aircraft, when flying at a small angle of attack, there is a resistance to roll because the angle of attack, and so the lift, will increase on the wing going down and decrease on the wing going up. However, this correcting (or restoring) effect will only last while the aircraft is actually rolling. It should be noted that this only happens when the angle of attack is small. If the angle of attack is large and near the stalling angle, then the increased angle on the stalling wing may cause a decrease in lift, and the decreased angle on the other side will result in an increase of lift. Thus, the new forces will tend to roll the aircraft still further, leading to autorotation.

10.4.1   Dihedral Angle

A common method of obtaining lateral stability is by the use of a dihedral angle on the main wings, illustrated in Figure 10.3. The dihedral angle is the angle between the wing and the horizontal. If the wings are inclined upwards towards the tips, the dihedral is positive. If the wings are inclined downwards, the angle is negative and termed *anhedral*.

When both wings are equally inclined, the resultant lift on the wings will be vertically upwards and will exactly balance the weight. However, if one wing becomes lower than the other, as shown in Figure 10.4, then the resultant lift on the wing will be slightly inclined in the direction of the lower wing while the weight will remain vertical.

Therefore, the two forces will not balance each other, and there will be a small resultant force acting in a sideway and downward direction. This force is temporarily unbalanced, and therefore the aircraft will move in the direction of this force – that is, the aircraft will *sideslip*, as illustrated in Figure 10.5. This will cause a flow of air in the opposite direction to the slip. This has the effect of increasing the angle of attack of the lower wing and decreasing that of the upper wing. The lower wing therefore would generate more lift, and a restoring moment would result. Also the

Figure 10.3   An aircraft with dihedral angle.

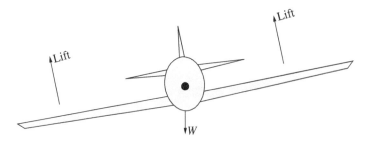

Figure 10.4   An aircraft with rolling tendency.

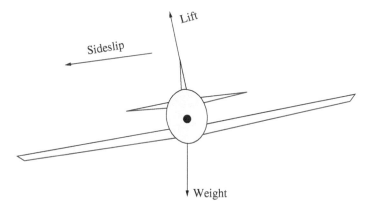

Figure 10.5 An aircraft sideslipping.

wing tip of the lower wing will become the leading edge during the sideslip. This will cause the centre of pressure across the chord to move nearer to the leading edge, so that centre of pressure distribution along the span will now be on the lower wing. Both these reasons will make the lower wing to generate more lift, and after a slight slip sideways, the aircraft will roll back in to its proper position. Because of the presence of the fuselage, the flow of air caused by the sideslip will not reach a larger portion of the raised wing. However, this effect depends very much on the position of the wing relative to the fuselage.

The leading edge effect on the lower wing and the shielding of the upper wing by the fuselage occur almost on all types of aircraft. Thus, an aircraft has a sufficient degree of lateral stability without any dihedral angle. This lateral stability will become too much if some of the following effects are present. Even if there is no actual dihedral angle on the wings, these methods of achieving lateral stability may be described as having a *dihedral effect*.

10.4.2 High Wing and Low Centre of Gravity

For aircraft with high wing, the lateral stability is high because of the low position of centre of gravity. When an aircraft sideslips, the lift of the lower wing becomes greater than the lift of the higher wing. Also, a small sideways drag force is introduced. The resultant force on the wing will be in the direction, as shown in Figure 10.6.

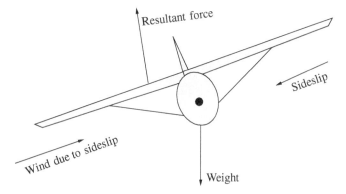

Figure 10.6 Resultant force on a high-wing aircraft in sideslip.

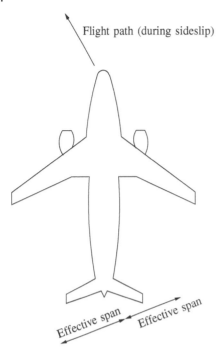

Flight path (during sideslip)

Figure 10.7   An aircraft with sweepback in sideslip.

Effective span   Effective span

It is seen that the resultant force does not pass through the centre of gravity, and therefore there will be a small moment that will tend to roll the aircraft back to a level condition. This will occur even on a low-wing aircraft but is more effective on a high-wing aircraft because the moment arm is greater. For this reason, a high-wing aircraft requires less dihedral than a low-wing aircraft.

10.4.3   Lateral Stability of Aircraft with Sweepback

The sweepback angle itself will promote lateral stability. For example, if the left wing drops, there will be a sideslip to the left, and the left wing, in effect, will have a higher aspect ratio than the right wing to the correcting airflow, as shown in Figure 10.7. It will therefore experience more lift, leading to the recovery of the aircraft.

10.4.4   Fin Area and Lateral Stability

The positioning of components such as the fuselage, fin and rudder, and wheels can have considerable influence on lateral stability. All these will present areas at right angles to any sideslip. Therefore, there will be pressure upon them. *If the pressure on these surfaces is high above the centre of gravity, the force due to these pressures will tend to restore* the aircraft to an even keel. This applies to many modern types that have a high tailplane, shown in Figure 10.8.

Such types may have anhedral on the main planes to counterbalance this effect and prevent too great a degree of lateral stability. However, if the side surfaces are low, the pressure on them will tend to roll the aircraft still more and so cause lateral instability, although this must be balanced against the effect on high wing compared with the centre of gravity position. The flow due to sideslip on a low-slung fuselage is shown in Figure 10.9.

From the discussions on sideslip, it is evident that whatever be the method of obtaining lateral stabilities, correction only takes place after a sideslip towards the low wing.

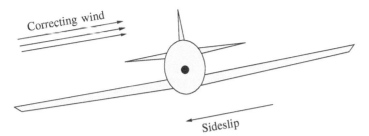

Figure 10.8 A high-fin aircraft in sideslip.

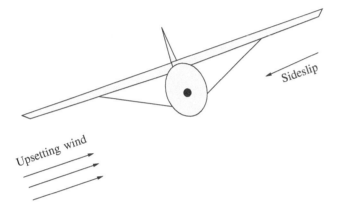

Figure 10.9 Flow due to sideslip on a low-slung fuselage.

10.5 Directional Stability

Directional stability of an aircraft is its ability to return to the original mode of flight, when it is disturbed from its course. For an aircraft in straight and level flight, the airflow will approach it directly from the front, that is, parallel to its longitudinal axis, as shown in Figure 10.10a. If it is deflected from its course, due to its momentum, it will for a short time tend to continue moving in its original direction. Therefore, the longitudinal axis will be inclined to the airflow, as illustrated in Figure 10.10b, and a pressure will be created on all the side surfaces on one side of the aircraft.

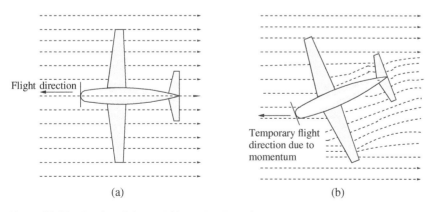

Figure 10.10 An aircraft in level flight (a) before disturbance and (b) after disturbance.

If the turning effect of the pressure behind the centre of gravity (*cg*) is greater than the turn effect in front of the *cg*, the aircraft will turn to its original course. On the other hand, if the turning effect in front of *cg* is greater than that behind the *cg*, the aircraft will turn still further off its course. Thus, the turning effect or the moment matters, and not the actual pressure. Therefore, it is not merely a question of how much side force, but also of the distance from the *cg* of each side surface. For example, a small fin at the end of a long fuselage may be as effective in producing directional stability as a large fin at the end of a short fuselage. Also, sometimes there may be more side surface in the front than in the rear, but rear surface will be at a greater distance. All the side surfaces of an aircraft, including that presented by wings with dihedral, affect the directional stability, but the *fin* is allotted the particular task of finally adjusting matters, and its area is settled accordingly.

There is a close resemblance between the directional stability of an aircraft and the action of a weathercock that always turns into the wind. However, it is essential to note that there are the following two distinct differences between an aircraft and a weathercock:

- An aircraft is free not only to yaw but also to move sideways.
- The wind in the case of an aircraft is not the wind due to the atmospheric motion but the wind caused by the motion of the aircraft through the air.

10.6 Lateral and Directional Stability

The sideslip, which is essentially the lateral stability, will cause an air pressure on the side surfaces that have been provided for directional stability. The effect of this pressure will be to turn the nose in to the relative wind, that is, in this case, towards the direction of sideslip. The aircraft, therefore, will turn off its original course and in the direction of the lower wing. It is to be noted that the greater the directional stability, the greater the tendency to turn off the course in a sideslip. This turn will cause the raised wing, now on the outside of the turn, to travel faster than the inner of lower wing and therefore to obtain more lift, causing the aircraft to bank still further. By this time the nose of the aircraft has probably dropped, and this tendency combines all three stabilities involved. The best way of seeing all these happen in real life is to watch a model aircraft flying in a gusty day. The light loading and slow speed of the model make it possible to watch each step in the proceedings, whereas in the full-scale aircraft it all happen more quickly, and also the pilot usually interferes by using his controls. For example, if the left wing drops and the pilot applies rudder so as to turn the aircraft to the right, he will probably prevent it from departing appreciably from its course.

If the aircraft has to be turned to the left, instead of applying rudder, simply bank the aircraft to the left. This banking will make the aircraft to slip inwards and turn off to the left. However, as far as the yaw is concerned, the rudder is still responsible. The difference is simply that the rudder and fin are brought into effect by the inward sideslip, instead of by application of rudder that tends to cause an outward skid. It should also be noted that although it may not be practical, and most sensible, to commence to turn in certain aircraft without application of rudder, such a turn cannot be absolutely perfect. There must be an inward sideslip.

If an aircraft is sideslipping and skidding, as a crab, lateral stability will come into play and cause the aircraft to roll away from the leading wing. Thus, a roll causes a yaw, and a yaw causes a roll, and the study of the two cannot be separated.

If the stability characteristics of an aircraft are such that it is very stable directionally and not very stable laterally, for example, if it has large fin and rudder and a little or no dihedral angle, or other 'dihedral effect', it will have a marked tendency to turn into a sideslip, and to bank at progressively steep angles, that it may get into an uncontrollable spiral – this is sometimes called *spiral instability*, but note that it is caused by too much of directional stability.

If, on the other hand, the aircraft is very stable laterally and not very stable directionally, it will sideslip without any marked tendency to get into sideslip. Such an aircraft is easily controllable by the rudder, and if the rudder only is used for a turn, the aircraft will bank and make quite a nice turn.

10.7 Control of an Aircraft

The control of an aircraft may broadly be classified as *longitudinal control, roll control*, and *directional control*. Longitudinal control is provided by the *elevators* or movement of the whole tailplane. Roll control is provided by the *ailerons*, and the directional control is provided by the *rudder*.

When a control surface is moved, it will alter the angle of attack and the camber of the complete surface and therefore change the force upon it, as illustrated in Figure 10.11.

The elevators and ailerons are controlled by movements of a *control columns* on which is mounted a handwheel. Pushing the control column forward lowers the elevators, thus increasing the lift on the tailplane and making the nose of the aircraft drop. Turning the handwheel counterclockwise lowers the right-hand aileron and raises the left-hand aileron, thus rolling the aircraft left wing down.

In aircraft with small cockpit, such as fighter aircraft, there is no handwheel, and instead the control column moves to left and right as well as backwards and forwards. A small version of this type of 'joystick' control is now used on many aircraft. It is placed to one side of the pilot and consequently referred to as a side-stick.

The rudder is controlled by foot pedals. Pushing the left pedal forward deflects the rudder to the left and therefore turns the nose of the aircraft to the left.

10.8 Balanced Control

Powered controls are used on larger aircraft, but on smaller aircraft manual controls are still used. The forces that the pilot has to exert to move the control are usually small, in normal flight. However, in bumpy weather, the continuous movement required becomes tiring during long flights, especially when the control surfaces are large and the speeds are fairly high. For this reason, controls are often balanced.

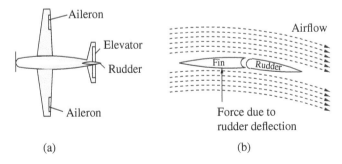

(a)   (b)

Figure 10.11   (a) Main control surface of aircraft. (b) Flow past a fin with rudder deflection.

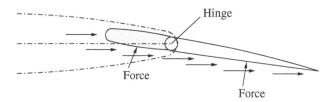

Figure 10.12   A hinged control surface.

Figure 10.13   (a) Horn balance and (b) inset hinge balance.

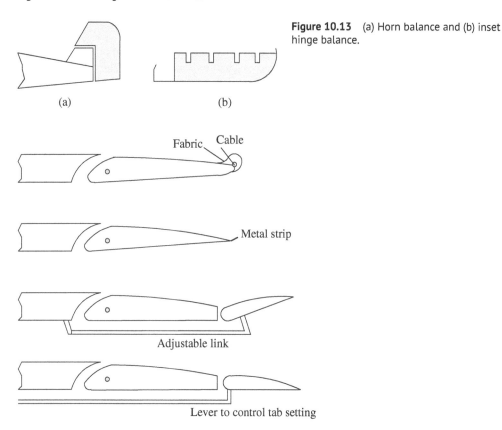

Figure 10.14   Movements of control tab.

A simple kind of aerodynamic balance is illustrated in Figure 10.12. The hinge is set back so that the air striking the surface in front of the hinge causes a force that tends to make the control move over still further; this balances the effect of the air that strikes the rear portion. This is effective but it must not be overdone. Overbalancing is dangerous since it may remove all feel of the control from the pilot. It should be noted that when the control surface is set at a small angle, the centre of pressure on the surface is well forward of the centre of the area, and at any angle if the centre of pressure is in front of the hinge, it will tend to take the control out of the pilot's hands or feet. Usually less than one-fifth of the surface may be in front of the hinge.

Horn balance and inset hinge balance, shown in Figure 10.13a,b, are two applications of hinge balance. In each of these, some part of the surface is in front of the hinge.

The servo type of balance shown in Figure 10.14 differs in principle. In this case, the pilot only moves the small extreme surface (in the opposite direction to normal), and, owing to the leverage, the force on the small surface helps to move main control in the required direction. The main attraction of the servo system of balance is that it was the forerunner of the *balance tabs* and *trimming tabs*. Even though the servo system suffered from many deflections, it did show how powerful is the effect of a small surface used to deflect the air in the opposite direction to that in which it is desired to move the control surface.

The next step was to apply this idea to an aileron when an aircraft was inclined to fly with one wing lower than the other. A strip of flexible model was attached to the trailing edge of the control surface and produced the necessary corrective force (bias). So far, the deflection of the air was in one direction, and so we obtained a force on the controls rather than a balancing system. The next step gave both balance and bias; the strip of metal became a *tab*, that is, an actual flap hinged to the control surface. The tab was connected by a link to a fixed surface (the tailplane, fin, or main plane), the length of this link being adjustable on the ground. When the main control surface moved in one direction, the tab moved in the other and thus experienced a force that tended to help the main surface to move – hence the balance. By adjusting the link, the tab could be set to give an initial force in one direction or the other – hence the bias.

A tab with a spring inserted between the tab and the main control system is called *spring tab*. The spring tab may be used to modify the system in the following two ways:

- The amount of tab movement decreases with speed, thus preventing the action being too violent at high speed.
- The tab does not operate until the main control surface has been moved through a certain angle or until a certain control force is exerted.

The final step in the balancing requires a little mechanical ingenuity, but otherwise it was a natural development. The pilot has the provision to adjust the bias while in the air and thus to correct any flying faults or out-of-balance effects as and when they occurred. On small aircraft with manual controls, these tabs may be fitted to all of the primary control surfaces. The pilot can adjust the

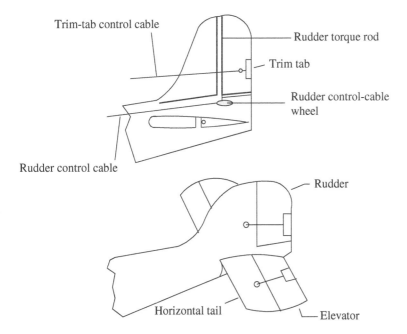

Figure 10.15 Location of control tab.

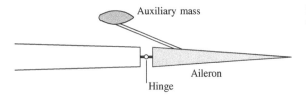

Figure 10.16 Mass balancing of an aircraft wing.

control settings from within the cockpit and can thereby arrange the trim so that the aircraft will fly 'hands off' in almost any flight condition. On aircraft with power-operated controls, such tabs are not required, and the trim wheels are simply used to reset the neutral or hands-off position of either the control column or the actuator system. The sequence of the evaluation of control tabs and their location on the aircraft structure are shown in Figures 10.14 and 10.15.

10.9 Mass Balance

Balancing of control surface is essential for the mechanical balancing of the mass of the control surface behind the hinge and for preventing the *flutter* that is liable to occur at high speed. Control surfaces are often balanced by fitting a mass in front of the hinge, as shown in Figure 10.16. Flutter is a vibration caused by the combined effect of the change in pressure distribution over the surface as the angle of attack is altered, and the elevator forces set up by the distortion of the structure itself. All structures are distorted when loads are applied. If the structure is elastic, as all good structures may be, it will tend to spring back as soon as the load is removed or changes its point of application.

An aircraft wing or fuselage can be distorted in two ways, by bending and twisting. Each distortion can result in an independent vibration. Like all vibrations, this flutter is liable to become dangerous if the two effects add up. The flutter may affect the control surfaces such as aileron or the main planes or both. This complicated problem of flutter can be prevented with a stiff structure and mass balance of the control surfaces. When the old types of aerodynamic balance were used, for example, the inset hinge or horn hinge, the mass could be concealed inside the forward portion of the control surface. However, when the tap type of balance is used alone, the mass must be placed on a special arm sticking, out of front of the control surface. However, the best way to tackle the flutter is by increasing the rigidity of the structure and control system components.

Large aircraft and military type invariably have powered controls, and they are much less sensitive to problem of flutter as the actuating system is very rigid.

It is essential to note that the *mass is not simply* a weight for the purpose of balancing the control surface statically, for example, to keep the aileron floating when the control mechanism is not connected. It may have its effect, but it also serves to alter the moments of inertia of the surface and thus alters the period of vibration and the liability to flutter. It may help to make this clear if we realise that mass balance is just as effective on a rudder, when the weight is not involved as on an elevator or aileron.

10.10 Control at Low Speeds

Control at low speeds is essentially controlling the aircraft flying at large angle of attack. Therefore, it is of no use to enable an aircraft to fly slowly unless it is ensured that the pilot will have adequate control over it. To gain an understanding of control at low speeds, let us assume that owing to engine failure, a pilot has to make a forced landing. In this situation any experienced pilot will tend to stall his aircraft in an attempt to reach a distant field or to climb over some obstacles. Now the use of slots or flaps may postpone the stall, which may help him to obtain lift at slow speeds, but they will not give him efficient control.

Owing to decreased speed of the airflow over all the control surfaces, the forces acting on them will be less, and they will be 'sloppy'. However, this is not all. For example, while flying near the stalling angle if the aircraft has to be turned to the left, the control column has to be moved to the left, and at the same time the rudder has to be moved to the left. During this control, the effect of the right aileron going down should increase the lift on the right wing, but in practice it may decrease it, since it may increase the angle of attack beyond that angle that gives the maximum lift. Consequently, it is certain that the drag will be considerably increased on the right wing; this will tend to pull the aircraft round to the right. This yawing effect, caused by the ailerons, is present at nearly all angles of attack, but it becomes marked near the stalling angle, hence termed *aileron drag*.

Meanwhile, the lift on wing may have either decreased or increased accordingly to the exact angle of attack, but in any case the change in lift will be small. The drag, on the other hand, will almost certainly have decreased as the aileron is moved upward. Thus, the net result of the attempt to turn to the left is that there may or may not be a slight tendency to roll in to the left bank required for an ordinary left-hand turn while the drag on the wings will produce a strong tendency to turn to the right that may completely overcome the rudder's effort in to opposite direction, as illustrated in Figure 10.17.

The solution to overcome this problem is the following. When the stalling angle is reached, it shall be ensured that the movement of the controls will cause the same effect on the aircraft as in normal flight. The following improvements will help to attain these effects overcome this problem.

- Increased turning effect from the rudder.
- Down-going aileron should not increase the drag.
- Up-going aileron should increase the drag.
- Down-going aileron should increase the lift at all angles.
- Up-going aileron should cause a loss of lift at all angles.

A large number of devices have been tried out in the attempt to satisfy these conditions. Even though most of them are partially successful, none of them has solved the problem completely.

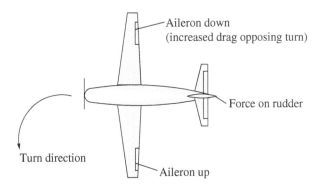

Figure 10.17 Forces on an aircraft during a turn at large angle of attack.

Let us review some of them and assess their success:

- Use of very large rudder with sufficient power to counter the yawing effect of the ailerons in the wrong direction. The disadvantage of this is the size of the rudder required to obtain the desired result is excessive for normal flight.
- A *washout* or decrease of angle of incidence, towards the wing tips. This means that when the centre portions of the wings are at their stalling angle, the outer portions are well below the angle, and therefore the aileron will function in the normal way. The disadvantage of this arrangement is that the washout must be considerable to have any appreciable effect on the control, and the result will be a corresponding loss of lift from the outer portion of the wing in normal flight.
- *Fraise* or other specially shaped ailerons, illustrated in Figure 10.18. This aileron when moved downwards the complete top surface of the wing and the aileron will have a smooth, uninterrupted contours causing very little drag. However, when the aileron is moved upwards, it will project below the bottom surface of the wing, as shown in Figure 10.18, and cause excessive drag.
- *Differential aileron*, shown in Figure 10.19, is a simple device. Instead of the two ailerons moving equally up or down, a simple mechanical arrangement of the controls causes the aileron that moves upwards to move through a larger angle than the aileron that moves downwards. The idea here is to increase the drag and decrease the lift on the wing with up-going aileron, while at the same time the down-going aileron, owing to its smaller movement, will not cause excessive drag.
- *Slot-cum-aileron* control is shown in Figure 10.20. The slot may be of automatic type, or the slot may be interconnected to the aileron in such a way that when the aileron is lowered, the slot is opened, while when the aileron is raised, or in its neutral position, the slot is closed. By this means, the down-going aileron will certainly serve to increase the lift for several degrees beyond the stalling angle, nor will the drag on this wing become very large since the open slot will lessen the formation of eddies. We shall therefore obtain a greater tendency to roll in the right direction and less tendency to yaw in the wrong direction. This is exactly what is required, and the system proved to be very effective in practice.

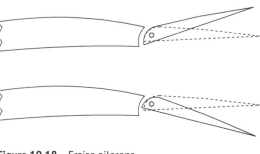

Figure 10.18 Fraise ailerons.

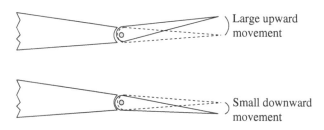

Figure 10.19 Differential ailerons.

Figure 10.20 Slot-cum-aileron control.

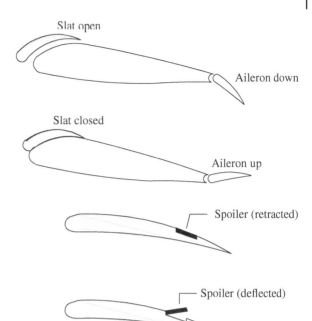

Slat open

Aileron down

Slat closed

Aileron up

Figure 10.21 Spoiler.

Spoiler (retracted)

Spoiler (deflected)

- *Spoiler control*: Spoilers are long narrow solid strips normally fitted to the upper surface of the wing, as shown in Figure 10.21. In normal mode of flight, they lie flush with the surface and have no effect on the performance of the wing. However, the spoilers can be connected to the aileron controls in such a way that when an aileron is moved up beyond a certain angle, the spoiler is raised at a large angle to the airflow or comes up through the slit, causing turbulence, decrease in lift, and increase in drag. This means that the wing on which the aileron goes down gets more lift, and very little extra drag, while on the other wing the lift is 'spoiled' and the drag greatly increased. Thus, we have a large rolling effect in the right direction combined with a yawing effect, also in the right direction.

The advantage is that the mechanical operation of the spoiler is easy, since the forces upon it are small. The spoilers also serve as an effective break. It should be noted that the spoilers are used as an aid to lateral control at low speeds. However, in the case of high-speed aircraft, the spoiler control may result in undesirable characteristics caused by compressibility.

10.11 Power Controls

All big aircraft nowadays are filled with power-actuated control surfaces that are very easy to operate even on larger airplanes. Because such controls offer virtually no natural resistance, they are given some form of artificial *feel*, a resistance that is designed to increase with flight speed so that the control system feels like a direct mechanical linkage.

The control column often incorporates a 'stick shaker' that operates when the aircraft approaches a stalled condition. This reproduces the shaking that normally occurs on simple mechanical systems due to the buffeting of the control surfaces caused by turbulence. With advanced and reliable electronic devices, it is now possible to make control systems of immense complexity that respond smoothly over a wide range of flight conditions and contain many built-in safety features. To prevent loss of control in the event of power failure, the systems are usually duplicated, triplicated, or even quadruplicated.

10.12 Dynamic Instability

The instabilities in the form of oscillations or deviations from the desired fight path that vary with time are called *dynamic instabilities*. For example, spiral instability is a dynamic instability.

10.13 Summary

The ability of the aircraft to return to the same flying mode, when slightly disturbed from that condition, without any effort on the part of the pilot is called *stability*.

Stability is different from balance or *trim* of an aircraft.

An aircraft that tends to return to its original position, when disturbed, is said to be stable. An aircraft that tends to move further away from the original position, when disturbed, is said to be *unstable*. However, if the disturbance sets the aircraft in a new position, the aircraft is said to have *neutral stability*.

The degree of stability would differ according to the *stick-fixed* and *stick-free* conditions.

A factor that influences the stability is the *power-off* or *power-on* condition.

The word 'control' means the power of the pilot to manoeuvre the aircraft into any desired position.

The stability or control of an aircraft concerning *pitching* about the lateral axis is called *longitudinal stability* or *control*, respectively. Stability or control concerning rolling about longitudinal axis is called *lateral stability* or *control*, respectively. Finally, stability or control concerning yawing about the normal axis is called *directional stability* or *control*.

To obtain stability in pitching, it is essential to ensure that if the angle of attack is increased, forces will act in such a way as to depress the nose, and this decreases the angle of attack.

The angle between the tailplane chord and wing chord is called *longitudinal dihedral*. Longitudinal dihedral is regarded as an important characteristic dictating the longitudinal stability of most type of aircraft.

When the tailplane is in front of the wings, the tail must have greater angle than the wings to have longitudinal dihedral.

For lateral stability, the aircraft must have the capability to restore its original mode of flight, when it is subjected to a slight roll.

A common method of obtaining lateral stability is by the use of a dihedral angle on the main wings. If the wings are inclined upwards towards the tips, the dihedral is positive. If the wings are inclined downwards, the angle is negative and termed *anhedral*.

Two forces will not balance each other, and there will be a small resultant force acting in a sideway and downward direction. This force is temporarily unbalanced, and therefore the aircraft will move in the direction of this force – that is, the aircraft will *sideslip*.

For aircraft with high wing, the lateral stability is high because of the low position of centre of gravity. When an aircraft sideslips, the lift of the lower wing becomes greater than the lift of the higher wing.

The sweepback angle itself will promote lateral stability.

The positioning of components such as the fuselage, fin and rudder, and wheels can have considerable influence on lateral stability. If the pressure on these surfaces are high above the centre of gravity, the force due to these pressures will tend to restore the aircraft to an even keel.

Directional stability of an aircraft is its ability to return to the original mode of flight, when it is disturbed from its course.

If the turning effect of the pressure behind the *cg* is greater than the turn effect in front of the *cg*, the aircraft will turn to its original course. On the other hand, if the turning effect in front of *cg* is greater than that behind the *cg*, the aircraft will turn still further off its course.

The sideslip, which is essentially the lateral stability, will cause an air pressure on the side surfaces that have been provided for directional stability. The effect of this pressure will be to turn the nose in to the relative wind, that is, in this case, towards the direction of sideslip.

If the aircraft has to be turned to the left, instead of applying rudder, simply bank the aircraft to the left. This banking will make the aircraft to slip inwards and turn off to the left.

If an aircraft is sideslipping and skidding, as a crab, lateral stability will come into play and cause the aircraft to roll away from the leading wing. Thus a roll causes a yaw, and a yaw causes a roll, and the study of the two cannot be separated.

If an aircraft is very stable laterally and not very stable directionally, it will sideslip without any marked tendency to get into sideslip.

The control of an aircraft may broadly be classified as *longitudinal control, roll control*, and *directional control*. Longitudinal control is provided by the *elevators* or movement of the whole tailplane. Roll control is provided by the *ailerons*, and the directional control is provided by the *rudder*.

The elevators and ailerons are controlled by movements of a *control columns* on which is mounted a hand-wheel.

The rudder is controlled by foot pedals. Pushing the left pedal forward deflects the rudder to the left and therefore turns the nose of the aircraft to the left.

Powered controls are used on larger aircraft, but on smaller aircraft manual controls are still used.

Balancing of control surface is essential for the mechanical balancing of the mass of the control surface behind the hinge and for preventing the *flutter* that is liable to occur at high speed. Control surfaces are often balanced by fitting a mass in front of the hinge.

An aircraft wing or fuselage can be distorted in two ways, by bending and twisting. Each distortion can result in an independent vibration. Like all vibrations, this flutter is liable to become dangerous if the two effects add up. The flutter may affect the control surfaces such as aileron or the main planes or both.

Large aircraft and military type invariably have powered controls, and they are much less sensitive to problem of flutter as the actuating system is very rigid.

Control at low speeds is essentially controlling the aircraft flying at large angle of attack.

All big aircraft nowadays are filled with power-actuated control surfaces that are very easy to operate even on larger airplanes.

The instabilities in the form of oscillations or deviations from the desired fight path that vary with time are called *dynamic instabilities*. For example, spiral instability is a dynamic instability.

11

Manoeuvres

11.1 Introduction

An aircraft has six degrees of freedom: three translational and three rotational along and about the longitudinal, lateral, and normal axes, as illustrated in Figure 11.1. The *longitudinal axis* is a straight line running from the foremost point to the rearmost point and passing through the centre of gravity and is horizontal when the aircraft is in level flight, as shown in Figure 11.1a. The aircraft may travel forward or backward along this axis. Backward motion – such as a tail side – is one of the rare of all manoeuvres. The forward motion along the longitudinal axis is the most common of all manoeuvres and is the main feature of *straight and level flight*. Any rotary motion about the longitudinal axis is called *rolling*.

The lateral axis is a straight line passing through the centre of gravity and at right angles to both longitudinal and normal axes, as shown in Figure 11.1b. It is horizontal when the aircraft is in rigging position and parallel to the line joining the wing tips. The aircraft may travel to right or left along the lateral axis; such motion is called *sideslipping* or *skidding*. Rotary motion of the aircraft about the lateral axis is called *pitching*.

The normal axis is a straight line passing through the centre of gravity and vertical when the aircraft is in rigging position and is at right angle to the longitudinal axis, as shown in Figure 11.1c. The aircraft may travel upwards or downwards along this axis, as in *climbing* or *descending*. Rotary motion of the aircraft about the normal axis is called *yawing*.

The longitudinal, lateral, and normal axes must be considered as moving with the aircraft and always remaining *fixed relative* to the aircraft, for example, the lateral axis will remain parallel to the line joining the wing tips in whatever attitude the aircraft may be.

The manoeuvres of an aircraft are the following:

- Movement forward or backwards.
- Movement up or down.
- Movement sideways, to right or left.
- Rolling.
- Yawing.
- Pitching.

Introduction to Aerospace Engineering: Basic Principles of Flight, First Edition. Ethirajan Rathakrishnan.
© 2021 John Wiley & Sons, Inc. Published 2021 by John Wiley & Sons, Inc.
Companion Website: www.wiley.com/go/Rathakrishnan/IntroductiontoAerospaceEngineering

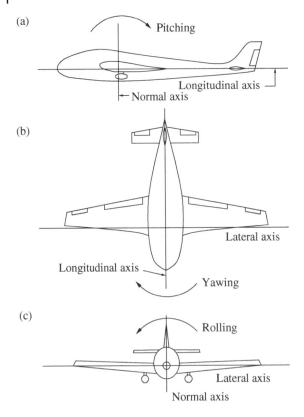

(a)

(b)

(c)

Figure 11.1 Illustration of translational and rotational motions of an aircraft.

11.2 Acceleration

The acceleration of an aircraft is the greatest during take-off. Also, the acceleration of an aircraft along its line of flight is relatively unimportant. However, the acceleration due to the change in direction of flight is of utmost importance.

We know that, for a body to move on a curved path, it is necessary to supply a force towards the centre of the curved path, this force being directly proportional to the acceleration required. Such a force is called the *centripetal force*. The body will cause a reaction on whatever makes its travel on a curved path. This reaction is termed *centrifugal force*.

For an aircraft travelling at a velocity V on a circumference of a circular path of radius r, the acceleration towards the centre of the circular path is V^2/r. If the mass of the aircraft is m, the centripetal (or centrifugal) force, F, becomes

$$F_c = m \times \text{acceleration}$$

that is,

$$F_c = m \, \frac{V^2}{r} \text{ newton} \qquad (11.1)$$

The acceleration being V^2/r shows that the acceleration is dictated by the flight speed, V, or flight path radius, r. However, the velocity has a greater effect on acceleration than the radius. This implies that at high speed, tight turns at small radius need a large force towards the centre of the curve.

At this stage it is important to understand the meaning of the vocabulary commonly used by the pilots, namely, 'g' factor. Pilots often talk about pulling a certain number of g. This quantity is just a number and not an acceleration. It has no units, and it simply represents a *factor*, which *when multiplied by the weight gives the total force that must be applied to a body to balance the combined effects of gravity and centripetal acceleration.* It is really a *load factor*, because it reflects the loads and stresses in the airframe increase during a manoeuvre.

Now let us consider the pull-out of a dive of an aircraft with a mass of 1000 kg, when it is subjected to a centripetal acceleration of 3g at the bottom of the manoeuvre. The *centripetal force* required to provide this acceleration will be $1000 \times 3g$ newtons. This force of 3000g newtons is a force equal to three times the weight. This centripetal force should be provided by the lift from the wings. However, the wing has to support the weight of the aircraft of 1000g newtons also. Thus, the wing has to support (3000g + 1000g) 4000g, which is four times the weight. The pilot therefore refers to this as a 4'g' manoeuvre. This 4'g' pull-out in this problem involves 3g centripetal acceleration.

Now think about what happens if the aircraft is at the top of a loop at the same speed and with the same radius of curvature. At the top of the loop, the centripetal acceleration required will be 3g, and the total force required to produce this will still be 3000g newtons. However, the weight of the aircraft will provide part of the centripetal force required (1000g newtons). Thus, the wing lift (now pulling downward) required is only 2000g newtons. That is, even though the centripetal acceleration is 3g, wings only have to provide twice as much lift force as in level flight: (2× the weight). Also, the pilot will be squashed so firmly into his seat at the bottom of the pull-out manoeuvre as his weight is trying to pull him out of the seat. Because the lift required is only (2× the weight), the load factor is only 2, and the pilot would call this as 2g manoeuvre.

If the loop was performed at the same speed, but with three times the radius, then the centripetal acceleration would be 1g, so that the centripetal force required becomes (1× weight), and the weight alone could provide all the centripetal force required. The wings need to produce no lift at all. This is therefore called a *zero* 'g' manoeuvre. The pilot would feel weightless.

11.3 Pulling Out from a Dive

For pulling out from a dive, the wings will have to generate more lift to provide the necessary centripetal acceleration. This means that the bending stress on the wing will increase. For example, in a 3g pull-out, the stress acting on the wing will be three times the stress during normal level flight. Design and safety requirements will specify the maximum load factor that the aircraft must be able to withstand, and all the stresses that would act on the aircraft have to be determined for this condition. Aircraft that are fully aerobatic type have high specified maximum load factor, whereas civil transports have a lower load factor.

To meet the lift increase required during a pull-out, the lift coefficient C_L should be increased by increasing the angle of attack. We know that there is a limiting angle of attack where the maximum C_L is reached, and any attempt to increase the angle of attack beyond this level will result in the aircraft stalling. The greater the centripetal acceleration required, the higher the stalling speed. Stalling while attempting to pull out too steeply is a condition that pilots must avoid at all times. Typical load factor for an aircraft having a level flight stalling speed of 30 m/s, pulling out of a steep dive, is illustrated in Figure 11.2, showing the load factor (g level), the stalling speed, and the centripetal acceleration.

Apart from the loads on the airframe, any manoeuvre involving large centripetal accelerations will have a physical effect on the pilot. The pilot's head will feel heavy, and he will experience

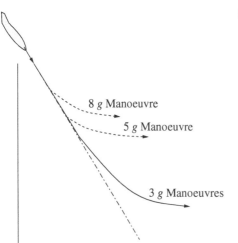

Figure 11.2 Pulling out of a dive.

8 *g* Manoeuvre

5 *g* Manoeuvre

3 *g* Manoeuvres

difficulty in moving his arms, which now feel several times heavier than normal. Even at 1.5g, writing becomes difficult. Also the centripetal acceleration can affect the blood circulation. At around 4 − 5g, the heart will start to have difficulty in pumping blood to the head, and if this is too severe, everything will appear to turn grey at first, leading to the danger of 'blacking out' and losing consciousness.

Apart from the problem of aircraft stalling, the physiological factors also impose limit on the severity of the manoeuvre that can be performed. In fighter aircraft, several means are employed to increase the amount of centripetal acceleration that can be tolerated. One simple approach is to have the pilot lying as near the horizontal as possible while still being able to see where he is going. Another approach involves the use of special 'g' suits that inflate at strategic points to temporarily restrict the flow of blood from the head.

11.3.1 The Load Factor

To meet the extra loads likely to be encountered during aerobatics, every part of the aircraft is given a *load factor*, which varies according to conditions. This load factor range is from 4 to 8. That is, the various parts of the aircraft are made to have strength 4–8 times higher than that they need to have during straight and level flight.

11.3.2 Turning

During a turn, the inward centripetal force is provided by the *banking* of the aircraft so that the total lift on the wings, in addition to supporting the weight of the aircraft, can supply a force component towards the centre of turn, as illustrated in Figure 11.3.

Assume an aircraft of weight W newtons flying with velocity V m/s on a circumference of a circle of radius r metres. The acceleration towards the centre of the circle is V^2/r m/s^2. Therefore, the force required towards the circle is

$$F = \frac{W V^2}{g r} \text{ newtons}$$

If the wings of the aircraft banks at an angle of θ to the horizontal, and if this angle is such that the aircraft has no tendency to slip either inwards or outwards, then the lift L newtons will act at right

Figure 11.3 Forces acting on an aircraft in turn.

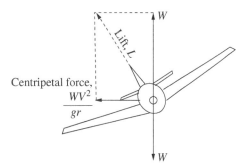

angles to the wings, and it must provide a *vertical component*, equal to W newtons, to balance the *weight*, and an *inward component*, equal to $W\,V^2/(g\,r)$ newtons, *to provide the acceleration towards the centre*. Thus,

$$\tan\theta = \frac{W\,V^2/(g\,r)}{W} = \frac{V^2}{g\,r} \tag{11.2}$$

This equation shows that there is a specified bank angle, θ, for any turn of radius r metres at a velocity of V m/s and the angle of bank is independent of the weight of the aircraft.

From Eq. (11.2), it is seen that an increase in velocity will result in increase in the bank angle and increase of turn radius will result in decrease of bank angle.

Example 11.1
If the bank angle of an aircraft flying at 120 m/s is 40°, find the radius of turn.

Solution
Given: $V = 120$ m/s and $\theta = 40°$. From Eq. (11.2), the radius of turn is

$$r = \frac{V^2}{g\,\tan\,\theta}$$

$$= \frac{120^2}{9.81 \times \tan\,40°}$$

$$= \boxed{1749.4 \text{ m}}$$

11.3.3 Loads During a Turn

During the turn, the lift on the wing is greater than during straight level flight. Also, the lift increases considerably with the angle of bank. This implies that the structural components, such as the wing spars, will have to carry considerably greater load than those of straight flight.

As seen in Figure 11.3, the lift during the turn is given by

$$L = \frac{W}{\cos\,\theta}$$

Thus, at 45° bank angle, $L = 1.414\,W$, and at 60° bank angle, $L = 2\,W$. These values of lift mean that at these angles of bank, the loads on the wing structures are 1.414 and 2 times, respectively, the loads of normal flight.

Whatever be the angle of bank, the lift on the wing must be provided by $\frac{1}{2}\rho V^2 S\,C_L$. It follows that $\frac{1}{2}\rho V^2 S\,C_L$ must be greater during a turn than during a normal flight, and this must be the *stalling speed*; the speed at the maximum value of C_L must go up in turn. At steep angles of bank, not only the stalling speed increases considerably, but also the other problems such as blacking out and possibility of structural failure of the aircraft are encountered. At this stage it is essential to note that steep turns can only be accomplished if the engine is powerful enough to keep the aircraft travelling at high speeds and at large angles of attack.

11.4 Correct Angles of Bank

A bank angle during which the wind will come straight ahead is called *correct bank angle*. If the bank is too much, the aircraft will sideslip inwards, and the flow over the aircraft will come from the inside of the turn. If the bank is too small, the wind will come from the outside of the turn, due to an outward skid on the part of the aircraft.

An indication of the nature of the bank would be a *plumb bob* hung in the cockpit out of contact with the wind. In normal flight this would hang vertically. During a correct bank it would not hang vertically, but in exactly the same position relative to the aircraft as it would in normal flight, that is, it would bank with the aircraft. If overbanked, the plumb line would be inclined inwards and, if underbanked, outwards from the above position. This plumb bob idea, in the form of pendulum, forms the basis of the sideslip indicator that is provided by the top pointer of the so-called *turn and bank indicator*. The pointer is geared so as to move in such a way that the pilot must move the control column *away from* the direction of the pointer, this being the instinctive reactions.

During a correct bank, the pilot will sit on his seat without any feeling of sliding either inwards or outwards; in fact, he will be sitting lighter in his seat than ever, his effective weight being magnified in the same proportions as the lift. If the aircraft overbanks, the pilot will tend to slide inwards. During underbanking, the pilot will tend to slide outwards.

11.5 Other Problems of Turning

To turn the aircraft the ailerons will be deflected. Once the turn is commenced, the outer wing will be travelling faster than the inner wing and will therefore experience more lift. To counter this differential lift acting on the outer and inner wings, it is not only necessary to take off the aileron control, but opposite ailerons should be applied by moving the control column against the direction of bank. This action is called *holding off bank*.

The effect of holding off bank is different in turns on a glide and on a climb. On a *gliding turn*, the whole aircraft will move the same distance downwards during one complete turn, but the inner wing, because it is turning on a smaller radius, will have descended on a steeper spiral than the outer wing. Therefore, the air will have to come up to meet it at a steeper angle; in other words the inner wing will have a larger angle of attack and so generate more lift than the outer wing. The extra lift obtained in this way may compensate the lift generated by the outer wing due to the increase in velocity. Thus, in a gliding turn there may be no need to hold off bank.

If an aircraft turns while *climbing*, the inner wing still describes a steeper spiral, but this time it is an upward spiral, so that air comes down to meet the inner wing more than the outer wing. This mode *reduces the angle of attack on the inner wing*. In this mode, the outer wing has more lift due to the combined effect of velocity and increased angle. This makes holding off bank during climb more essential than during a normal turn.

Another way of looking at the problem of gliding and climbing turn is to analyse the motion of an aircraft around its three axes during such turns. In a flat turn, a level turn without bank,

Figure 11.4 An aircraft in climbing turn.

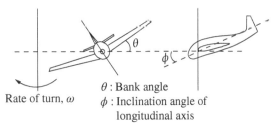

Rate of turn, ω

θ : Bank angle
ϕ : Inclination angle of
longitudinal axis

the aircraft experiences yaw only. However, in a banked level turn, the aircraft yaws and pitches. Turing of aircraft during a glide or climb will cause the aircraft to pitch, yaw, and roll. The aircraft rolls inwards in a gliding turn and rolls outwards during climbing turn. The inward roll results in extra angle of attack on the inner wing, and the outward roll results in extra angle of attack on the outer wing.

Examine the climbing turn shown in Figure 11.4 in terms of the rate of turn of the aircraft, ω; the bank angle, θ; and the pitch angle, ϕ.

In Figure 11.4, it is seen that

$$\text{Rate of yaw} = \omega \ \cos \ \phi \ \cos \ \theta$$
$$\text{Rate of pitch} = \omega \ \cos \ \phi \ \sin \ \theta$$
$$\text{Rate of roll} = \omega \ \sin \ \phi$$

For an aircraft in level flight, $\theta = 0$ and $\phi = 0$. Thus,

$$\text{Rate of yaw} = \omega$$
$$\text{Rate of pitch} = 0$$
$$\text{Rate of roll} = 0$$

For an aircraft in vertical bank or vertical pitch, $\theta = 90°$ and $\phi = 90°$; thus

$$\text{Rate of yaw} = 0$$
$$\text{Rate of pitch} = 0$$
$$\text{Rate of roll} = \omega$$

That is, during a vertical bank, there is *no roll* even though the aircraft may be descending during climbing.

Note that the rate of roll depends only on the angle of pitch, ϕ. That is, for $\phi = 0$, there is no pitch or roll even when the aircraft descends or climbs. At large angles of bank, θ, there is only a small difference in velocity, and angle, between inner and outer wings. This renders the need for holding off bank less important, but there are other problems associated with bank at large θ.

A turn involving pure yaw is called *flat turn*, which is useful for a bomber aircraft approaching a target. However, in practice flat turn is difficult because of the following:

- The extra velocity of the outer wing will tend to bank the aircraft.
- The lateral stability acts in such a way as to try to prevent the outward skid by the banking aircraft.
- The side area is often insufficient to provide enough inward force to cause a turn except on a very large radius.
- The directional stability will oppose the action of the rudder and tend to put the nose of the aircraft back so that it will continue on a straight path.

11.6   Steep Bank

As the aircraft banks steeper, *the rudder gradually will take the place of elevators, and vice versa.* However, this has to be handled with caution, because in a vertical bank, for instance, the rudder will not be powerful in raising or lowering the nose as the elevators in horizontal flight. Therefore, a vertical bank without sideslip is impossible, since in such a bank the lift will be horizontal and will provide no contribution towards lifting the weight. If such a bank has to be executed, then a straight upward inclination of the fuselage together with the propeller thrust provides sufficient lift.

This is applied only to a continuous vertical bank in which no height to be lost. During this turn, for a few moments, the bank is vertical. In general, the radius of turn can be reduced as the angle of bank is increased. However, even with a vertical bank, there is a limit to the smallness of the radius, because, quite apart from the question of sideslipping, the lift on the wings must provide all forces towards the centre. Thus, the lift is balanced by this force, that is,

$$\frac{1}{2}\rho V^2 S\, C_L = \frac{W\, V^2}{g\, r}$$

where ρ is air density, V is flight speed, S is the wing area, C_L is the lift coefficient, W is the weight of the aircraft, g is gravitational acceleration, and r is the radius of turn. Thus, the radius of turn becomes

$$r = \frac{2\, W}{\rho\, S\, g\, C_L} \tag{11.3}$$

In level flight the stalling speed, V_s, is given by

$$W = L = \frac{1}{2}\rho V_s^2 S\, C_{L_{\max}} \tag{11.4}$$

Substituting this into Eq. (11.3), we get

$$r = \frac{2\,\frac{1}{2}\rho V_s^2 S\, C_{L_{\max}}}{\rho\, S\, g\, C_L}$$

or

$$\boxed{r = \frac{V_s^2}{g}\,\frac{C_{L_{\max}}}{C_L}} \tag{11.5a}$$

This shows that the radius of turn will be least when C_L is equal to $C_{L_{\max}}$. That is, when the angle of attack is the stalling angle, the radius of turn becomes

$$r = \frac{V_s^2}{g} \tag{11.5b}$$

Note that the minimum radius of turn is independent of the actual speed during the vertical banks; it is dictated by the stalling speed of the aircraft. Equation (11.5b) applies to all steep turns. The lower the stalling speed, the tighter the turn that the aircraft can make.

11.7   Aerobatics

Flight modes such as loop, spin, roll, nosedive, upside down, sideslip, inverted spin, and inverted loop are called aerobatics. Aerobatics should be performed only with aircraft suitable for them. These flight modes provide good training for the accuracy and precision in manoeuvre, which are essential for combat flying.

(a)

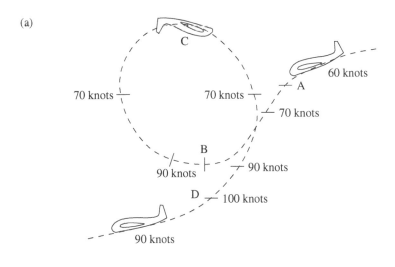

(b)

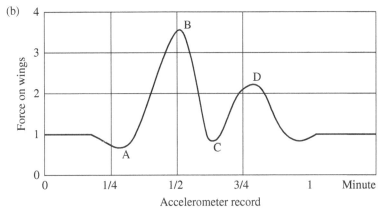

Figure 11.5 (a) An aircraft in loop and (b) the accelerometer diagram for the loop.

The movements of the aircraft during these aerobatics are complex and hence difficult to predict. Approximate path travelled by a slow aircraft during a loop and the corresponding accelerometer diagram showing the force variation on the wings during the manoeuvre are shown in Figure 11.5.

In Figure 11.5, it is seen that the greatest loads occur at the moment of entry, as in many manoeuvres. Also, at the top of the loop, the load is slightly less than normal, forcing the pilot to sit firmly on his seat in the upward direction. The load will still be in the same direction *relative to the aircraft* as in normal flight, and the plumb bob will be hanging upwards. Only in a bad loop the load at the top will become negative, causing the loads on the aircraft structure to be reversed.

A spin, illustrated in Figure 11.6, is an interesting manoeuvre. It differs from other manoeuvres in the fact that the wings are *stalled*, that is, are beyond the critical angle of attack, and this accounts for the lack of control which the pitot experiences over the movements of the aircraft while spinning. It is essentially a form of *autorotation*, which means that there is a natural tendency for the aircraft to rotate on its own accord.

In a spin the aircraft follows a steep spiral path, but the attitude while spinning may vary from the almost horizontal position to a *flat spin* to the almost vertical position of the *spinning nosedive*. That is, the spin, like a gliding turn or steep spiral, is composed of varying degrees of yaw, pitch,

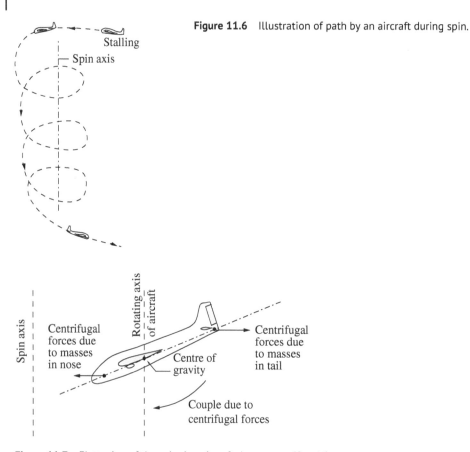

Figure 11.6 Illustration of path by an aircraft during spin.

Figure 11.7 Flattening of the spinning aircraft due to centrifugal force.

and roll. A flat spin is primarily yaw; a spinning nosedive is primarily roll. The amount of pitch depends on how much the wings are banked from the horizontal. In general, the air speed during a spin is comparatively low, and the rate of descent is also low. Any device, such as slots, which tend to prevent stalling, will also tend to minimise the danger of the accidental spin and may even make it impossible to carry out deliberately. The area and dispositions of the fin, rudder, and tailplane exert considerable influence on the susceptibility of the aircraft to spinning.

The way to get out of a spin is to come out of the stalled state by putting the aircraft in the nose-down and stop it from rotating by applying *opposite rudder*. In practice, the latter is usually done first because it is found that the elevators are not really effective until the rotation is stopped. The further back the centre of gravity and more masses that are disturbed along the length of the fuselage, the flatter and further the spin tend to become and more difficult to recover. This flattening of the spin is due to the centrifugal forces that act on the masses at the various parts of the aircraft, as illustrated in Figure 11.7.

During a roll the aircraft rotates laterally through 360°, but the actual path is in the nature of the horizontal corkscrew, these being a varying degrees of pitch and yaw, as illustrated in Figure 11.8.

In a slow roll the loads in the 180° position are reversed, as in inverted flight, whereas in the other extreme, the barrel roll, which is a cross between a roll and a loop, the loads are never reversed.

In a sideslip, there will be considerable wind pressure on all the side surfaces of the aircraft, notably the fuselage, the fin, and the rudder, while if the aircraft has a dihedral angle, the pressure

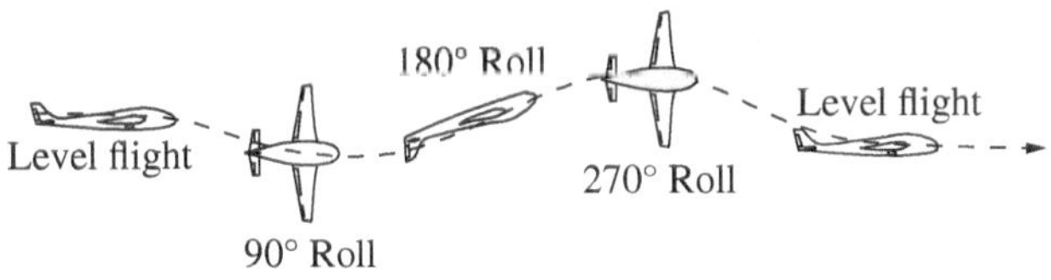

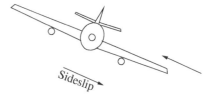

Level flight 90° Roll 180° Roll 270° Roll Level flight

Figure 11.8 Illustration of one full rotation of an aircraft in spin.

Figure 11.9 An aircraft in sideslip.

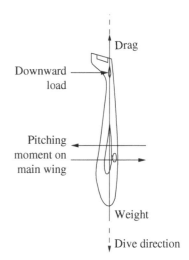

Sideslip

Figure 11.10 Forces on an aircraft during nosedive.

Drag

Downward load

Pitching moment on main wing

Weight

Dive direction

on the wings will tend to bring the aircraft on to even keel (a high-wing aircraft always has the tendency to turn the longitudinal axis of the aircraft into the relative wind, which is often referred to as the keel effect). A typical sideslip, showing the direction of the sideslip and the side wing, is illustrated in Figure 11.9. The sideslip is a useful manoeuvre for losing height just before landing. The small side area means that the aircraft drops very quickly if the sideslip is steep, and the directional stability is strong that it may be impossible to hold the nose of the aircraft up – by means of rudder. In addition, dropping of the nose causes even more increase in speed.

Nosedive is a form of gliding with gliding angle close to 90°. That is, it is a vertical descent. However, such deep dives are performed only rarely. In a vertical dive, an aircraft will eventually reach a steady velocity called the *terminal velocity*. In a vertical dive, the weight is entirely balanced by the drag, while the lift is zero. The angle of attack is very small or even negative. There is a large positive pressure near the leading edge on the top surface of the aerofoil, tending to turn the aircraft on to its back, and this is balanced by a 'down load' in the tailplane, as illustrated in Figure 11.10. Note that here 'down load' acts in the horizontal direction.

The terminal velocity of modern aircraft is very high, and it makes no difference whether the engine is running or not. The aircraft will lose so much height in attaining the terminal velocity that in practice, it is doubtful whether it can ever be reached. When the terminal velocity is near the speed of sound, the compressibility effects will be considerable. Thus, the compressibility effect is a special feature of nosediving. However, luckily as the aircraft goes near the earth, the terminal velocity will come down because of the greater density of atmosphere and increase in the speed of sound, owing to the higher temperature.

11.8 Inverted Manoeuvres

Inverted manoeuvres is essentially a flight mode in which the aircraft is in upside-down flight, as shown in Figure 11.11. To maintain the height during inverted flight, the engine must continue to run. The aircraft will be inverted; therefore, to produce an angle of attack, the fuselage will have to be in a 'tail-down' attitude. This attitude will affect the stability, although some aircraft have been more stable when upside down than the right way up, and considerable difficulty has been experienced in restoring the normal flight.

The *inverted spin* has most of its characteristics similar to the normal spin. However, the loads on the aircraft structure are reversed, and the pilot must rely on his straps to hold him in the seat.

The *inverted loop* or 'double bunt' is a manoeuvre in which the pilot is on the outside of the loop. The difficulty associated with inverted loop is due to the fact that in the normal loop, the climb to the top of the loop is completed while there is speed and power in hand and engine and aerofoils are functioning in the normal fashion, whereas in the inverted loop, the climb to the top is required during the second portion of the loop when the aerofoils are in the inefficient inverted position. The path traced by an aircraft in inverted loop is illustrated in Figure 11.12.

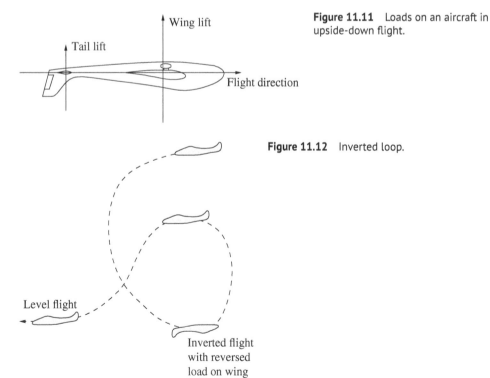

Figure 11.11 Loads on an aircraft in upside-down flight.

Figure 11.12 Inverted loop.

11.9   Abnormal Weather

Abnormal weather is one of the serious adverse effects during a flight. The unsteady conditions caused by abnormal weather condition such as turbulence can be quite considerable, as they must be reckoned with when designing commercial aircraft. The aircraft designed for performing aerobatics will be strong enough to withstand any loads caused by the adverse weather. The conditions that are likely to inflict the most adverse loads consist of strong gusty winds, hot sun, intermittent clouds, especially thunder clouds in which there is often considerable turbulence, and uneven ground condition.

11.10   Manoeuvrability

The inertia of an aircraft (or the moment of inertia of the various parts) will largely determine the ease of handling the aircraft during manoeuvres. The moment of inertia is the natural resistance of the aircraft to any form of rotation about the centre of gravity. A heavy mass that has a long distance away from a particular axis of rotation will make it more difficult to cause any rapid movement around the axis. Thus, masses such as engines far out on the wings result in a resistance to rolling about the longitudinal axis, and a long fuselage with large masses well forward or back will mean a resistance to pitching and yawing.

11.11   Summary

An aircraft has six degrees of freedom: three translational and three rotational along and about the longitudinal, lateral, and normal axes. The forward motion along the longitudinal axis is the most common of all manoeuvres and is the main feature of *straight and level flight*. Any rotary motion about the longitudinal axis is called *rolling*.

An aircraft may travel to right or left along the lateral axis; such motion is called *sideslipping* or *skidding*. Rotary motion of the aircraft about the lateral axis is called *pitching*.

An aircraft may travel upwards or downwards along this axis, as in *climbing* or *descending*. Rotary motion of the aircraft about the normal axis is called *yawing*.

It is important to understand the meaning of the vocabulary commonly used by the pilots, namely, 'g' factor. Pilots often talk about pulling a certain number of gs. This quantity is just a number and not an acceleration. It has no units, and it simply represents a *factor*, which *when multiplied by the weight gives the total force that must be applied to a body to balance the combined effects of gravity and centripetal acceleration*. It is really a *load factor*, because it reflects the loads and stresses in the airframe increase during a manoeuvre.

For pulling out from a dive, the wings will have to generate more lift to provide the necessary centripetal acceleration. To meet the lift increase required during a pull-out, the lift coefficient C_L should be increased by increasing the angle of attack.

Apart from the loads on the airframe, any manoeuvre involving large centripetal accelerations will have a physical effect on the pilot. The pilot's head will feel heavy, and he will experience difficulty in moving his arms, which now feel several times heavier than normal. Even at 1.5g, writing becomes difficult. Also the centripetal acceleration can affect the blood circulation. At around

$4 - 5g$, the heart will start to have difficulty in pumping blood to the head, and if this is too severe, everything will appear to turn grey at first, leading to danger of 'blacking out' and losing consciousness.

Apart from the problem of aircraft stalling, the physiological factors also impose limit on the severity of the manoeuvre that can be performed.

To meet the extra loads likely to be encountered during aerobatics, every part of the aircraft is given a *load factor*, which varies according to conditions. This load factor range is from 4 to 8.

During a turn, the inward centripetal force is provided by the *banking* of the aircraft so that the total lift on the wings, in addition to supporting the weight of the aircraft, can supply a force component towards the centre of turn.

During the turn, the lift on the wing is greater than during straight level flight.

A bank angle during which the wind will come straight ahead is called *correct bank angle*. An indication of the nature of the bank would be a *plumb bob* hung in the cockpit out of contact with the wind.

If an aircraft turns while *climbing*, the inner wing still describes a steeper spiral, but this time it is an upward spiral, so that air comes down to meet the inner wing more than the outer wing. This mode *reduces the angle of attack on the inner wing*. In this mode, the outer wing has more lift due to the combined effect of velocity and increased angle. This makes holding off bank during climb more essential than during a normal turn.

A turn involving pure yaw is called *flat turn*, is useful for a bomber aircraft approaching a target. As the aircraft banks steeper, *the rudder gradually will take the place of elevators, and vice versa*.

The radius of turn will be least when C_L is equal to $C_{L_{max}}$.

Flight modes such as loop, spin, roll, nosedive, upside down, sideslip, inverted spin, and inverted loop are called aerobatics. Aerobatics should be performed only with aircraft suitable for them.

A spin differs from other manoeuvres in the fact that the wings are *stalled*, that is, are beyond the critical angle of attack, and this accounts for the lack of control which the pitot experiences over the movements of the aircraft while spinning. It is essentially a form of *autorotation*, which means that there is a natural tendency for the aircraft to rotate on its own accord.

In a spin the aircraft follows a steep spiral path, but the attitude while spinning may vary from the almost horizontal position to a *flat spin* to the almost vertical position of the *spinning nosedive*.

The way to get out of a spin is to come out of the stalled state by putting the aircraft in the nose-down and stop it from rotating by applying *opposite rudder*.

During a roll the aircraft rotates laterally through 360°, but the actual path is in the nature of the horizontal corkscrew, these being a varying degrees of pitch and yaw.

Nosedive is a form of gliding with gliding angle close to 90°. That is, it is a vertical descent. However such deep dives are performed only rarely. In a vertical dive, an aircraft will eventually reach a steady velocity called the *terminal velocity*.

Inverted manoeuvres is essentially a flight mode in which the aircraft is in upside-down flight.

The *inverted spin* has most of its characteristics similar to the normal spin. However, the loads on the aircraft structure are reversed, and the pilot must rely on his straps to hold him in the seat.

The *inverted loop* or 'double bunt' is a manoeuvre in which the pilot is on the outside of the loop.

Abnormal weather is one of the serious adverse effects during a flight. The unsteady conditions caused by abnormal weather condition such as turbulence can be quite considerable.

The inertia of an aircraft (or the moment of inertia of the various parts) will largely determine the ease of handling the aircraft during manoeuvres.

12

Rockets

12.1 Introduction

A rocket engine is an engine that produces a force (a thrust) by creating a high-velocity output without using any of the constituents of the 'atmosphere' in which the rocket is operating. The thrust is produced because the exhaust from the rocket has a high velocity and therefore a high momentum. The rocket engine must, therefore, have to exert a force given by the product of the mass flow rate times the exhaust velocity, called the thrust, as illustrated in Figure 12.1.

The fact that the rocket engine does not use any constituent of the surrounding atmosphere means that it can operate in any part of the atmosphere and outside the atmosphere that makes it ideal for space propulsion. There are two basic types of rocket engine: chemical rockets and non-chemical rockets.

In a chemical rocket, a fuel and an oxidiser are usually supplied to the combustion chamber of the rocket. The chemical reaction between the fuel and the oxidiser produces high pressure and temperature in the combustion chamber, and the gaseous products of combustion can be expanded down to the ambient pressure, which is much lower than the combustion chamber pressure, giving a high velocity gaseous efflux from the rocket engine, as shown in Figure 12.2.

12.2 Chemical Rocket

There are two types of chemical rockets, *liquid-propellant rockets* and *solid-propellant rockets*. In a liquid-propellant rocket, the fuel and the oxidiser are stored in the rocket in liquid form and pumped into the combustion chamber. Arrangement of typical liquid-propellant rocket motor is shown in Figure 12.3.

In a solid-propellant rocket, the fuel and the oxidiser are in solid form, and they are usually mixed together to form the propellant. This propellant is carried within the combustion chamber. The arrangement of a solid propellant rocket is shown in Figure 12.4.

In a non-chemical rocket, the high efflux velocity from the rocket is generated without any chemical reaction taking place. For example, a gas could be heated to high pressure and temperature by passing it through a nuclear reactor, and it could then be expanded through a nozzle to give a high efflux velocity.

Introduction to Aerospace Engineering: Basic Principles of Flight, First Edition. Ethirajan Rathakrishnan.
© 2021 John Wiley & Sons, Inc. Published 2021 by John Wiley & Sons, Inc.
Companion Website: www.wiley.com/go/Rathakrishnan/IntroductiontoAerospaceEngineering

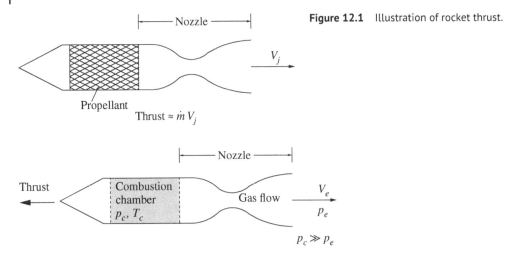

Figure 12.1 Illustration of rocket thrust.

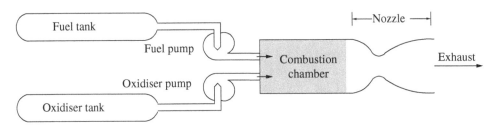

Figure 12.2 Thrust produced by the combustion gases.

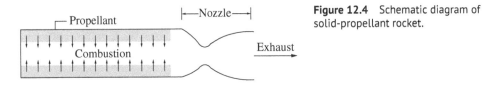

Figure 12.3 Schematic diagram of liquid-propellant rocket.

Figure 12.4 Schematic diagram of solid-propellant rocket.

The term rocket has frequently been used to describe both the thrust-producing device, that is, the engine, and the whole rocket-powered vehicle. To avoid confusion, especially in the case of large vehicles such as space launch vehicles, the propulsion device is now usually referred to as a rocket engine.

The advantages of liquid-fuelled rockets are the following:

1. Higher exhaust velocity (specific impulse).
2. Controllable thrust (throttle capability).
3. Restart capability.
4. Termination control.

The advantages of solid-fuelled rockets are the following:

1. Reliability (fewer moving parts).
2. Higher mass fractions (higher density implies lower tankage).
3. Operational simplicity.

Table 12.1 Some liquid and solid propellants and their specific impulse

Fuel	Oxidiser	I_{sp} (s)
Liquid propellants		
Hydrogen (LH$_2$)	Oxygen (LOX)	450
Kerosene	LOX	260
Monomethyl hydrazine (MMH)	Nitrogen tetroxide (N$_2$O$_4$)	310
Solid propellant		
Powdered Al	Ammonium perchlorate	270

Table 12.2 Rocket development events

Year	Event
300 BC	Gunpowder-filled bamboo tubes used for fireworks in China
1045	Military rockets in use in China
1895	Konstantin Tsiolkovsky derives the fundamental rocket equation
1926	Robert Goddard launches first liquid-fuelled rocket
1942	Wernher von Braun's team launches first successful A4 (V2)
1957	Sputnik launch
1958	Explorer I launch
1967	Saturn V first launch
1969	Apollo 11 Moon launch

Some of the widely used liquid and solid propellants and their specific impulse are listed in Table 12.1.

A brief survey of the rocket development events is listed in Table 12.2.

12.3 Engine Design

Chemical rocket engines combine knowledge of physics, chemistry, materials, heat transfer, and many other fields in a complicated, integrated system. The F-1 engine used in the first stage of the Saturn V rockets that launched the Apollo missions appears to be the first design.

The issues associated with the engine design are the following:

1. *Heat transfer*: This involves *radiative cooling*, radiating heat to space or conducting it to the atmosphere; *regenerative cooling*, running cold propellant through the engine before exhausting it; *boundary-layer cooling*, aiming some cool propellant at the combustion chamber walls; and *transpiration cooling*, diffusing coolant through porous walls.
2. *Nozzles*: *Rocket nozzles* are usually of an expansion–deflection design. This allows better handling of the transition from subsonic flow within the combustion chamber to supersonic flow as the propellant expands out at the end of the nozzle and produces thrust. Many nozzle variations exist. The governing equation for the magnitude of the thrust, in its simplest form, is

$$F = \dot{m} \, V_e + (p_e - p_a) A_e$$
$$= \dot{m} \, V_{eq}$$

where V_e is the exhaust velocity, $\dot{m}$ is the propellant mass flow rate, p_e is the pressure of the exhaust gases, p_a is the pressure of the atmosphere, A_e is the area of the nozzle at the exit, and V_{eq} is the equivalent exhaust velocity (that is, corrected for the pressure terms). If the pressure inside the chamber is too low, the flow will stagnate, while too high a pressure will give a turbulent exhaust – resulting in power wasted to transverse flow.

12.3.1 Saturn V

Wernher von Braun's team at Marshall Space Flight Center developed the three-stage Saturn V rocket. The Saturn V served as the workhorse of the Apollo Moon launches. Its first stage developed over 30 MN (7.5 million lbs) of thrust and burned about 14 tonnes of propellant per second for 2.5 minutes.

12.3.2 Space Shuttle Program

NASA's space shuttle fleet began setting records with its first launch on 12 April 1981 and continued to set high marks of achievement and endurance through 30 years of missions. Starting with Columbia and continuing with Challenger, Discovery, Atlantis, and Endeavour, the spacecraft has carried people into orbit repeatedly; launched, recovered, and repaired satellites; conducted cutting-edge research; and built the largest structure in space, the International Space Station. The final space shuttle mission, STS-135, ended 21 July 2011 when Atlantis rolled to a stop at its home port, NASA's Kennedy Space Center in Florida.

NASA is designing and building the capabilities to send humans to explore the solar system, working toward a goal of sending humans to a captured, relocated asteroid in the next decade and landing humans on Mars in the 2030s.

NASA is also making progress with the development of the Space Launch System (SLS) – an advanced heavy-lift rocket that will provide an entirely new capability for human exploration beyond Earth's orbit. SLS takes advantage of heritage hardware while also using modern manufacturing process such as friction-stir welding and adaptive manufacturing.

12.3.2.1 International Space Station

The International Space Station is the centrepiece of our human spaceflight activities in low-Earth orbit. The space station is fully staffed with a crew of six, and American astronauts will continue to live and work there in space 24 hours a day, 365 days a year. Part of the US portion of the station has been designated as a national laboratory, and NASA is committed to this unique resource for wide-ranging scientific research.

The space station is a test bed for exploration technologies such as autonomous refuelling of spacecraft, advanced life support systems, and human/robotic interfaces. US commercial companies working with NASA are well on their way to providing cargo and crew transportation to the orbiting laboratory, allowing the agency to focus its attention on sending astronauts on deep space missions, including asteroid and Mars.

12.4 Thrust Generation

Consider a control volume surrounding a rocket engine as shown in Figure 12.5.

The net force, F, exerted on the control volume must be equal to the rate of increase of momentum through the control volume. However, no momentum enters the control volume, so net force on control volume = rate at which momentum leaves control volume:

$$F = \dot{m}V_e \tag{12.1}$$

where

$$\dot{m} = -\frac{dm_f}{dt} \tag{12.2}$$

is rate at which mass leaves the control volume that, as indicated in Eq. (12.2), is equal to the rate at which the mass of propellant, m_f, is decreasing with time. V_e is the discharge velocity from the nozzle.

The force, F, on the control volume arises due to the pressure force exerted on the gases within the rocket engine and due to the difference in pressure over the surface of the control volume. Now, the pressure is the same everywhere on the surface of the control volume except on the nozzle exit plane that has area A_e. Hence, the net force on the control volume is given by

$$F = F_i - (p_e - p_a)A_e \tag{12.3}$$

However, the thrust is equal in magnitude to the force acting on the rocket, F_i, so the magnitude of the thrust is given by Eqs. (12.1) and (12.3) as

$$T = \dot{m}V_e + (p_e - p_a)A_e \tag{12.4}$$

Using Eq. (12.2), this can be expressed as

$$T = -\frac{dm_f}{dt}V_e + (p_e - p_a)A_e \tag{12.5}$$

The negative sign is due to the term dm_f/dt being negative, that is, the propellant mass m_f is decreasing.

Figure 12.5 Control volume surrounding a rocket engine.

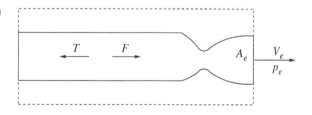

The last term in Eqs. (12.3)–(12.5) arises because of the difference between the pressure on the exhaust plane of the nozzle and the local ambient pressure. As discussed later in this section, the nozzle is usually designed so that $p_e = p_a$ under a chosen set of conditions that are usually termed the 'design conditions'. In most cases, the pressure term in the above equations is much smaller than the momentum term and can, therefore, often be neglected. In this case, the rocket thrust is given by

$$T = -\frac{dm_f}{dt}V_e \qquad (12.6)$$

12.5 Specific Impulse

Another term that is widely used in defining the performance of rocket engines is the specific impulse, I_{sp}. This can either be defined as the thrust per unit mass flow rate of propellant as

$$I_m = \frac{T}{\dot{m}} = \frac{T}{-dm_f/dt} \qquad (12.7a)$$

or it can be defined as the thrust per unit weight flow of propellant as

$$I_w = \frac{T}{\dot{W}} = \frac{T}{\dot{m}g} = \frac{I_m}{g} \qquad (12.7b)$$

The specific impulse is mainly dependent on the type of propellant used. An important parameter used in defining the overall performance of a propellant is the total impulse, I_T, which is equal to integral of thrust over the time of operation of the engine, t, that is,

$$I_T = \int_0^t T \, dt = -\int_0^t I_m \frac{dm_f}{dt} \, dt \qquad (12.8)$$

If the thrust is constant, the total impulse is given by

$$I_T = T \, t = -I_m \frac{dm_f}{dt} \, t = I_m m_{f0} = \frac{I_w m_{f0}}{g} \qquad (12.9)$$

where m_{f0} is the initial mass of propellant.

Example 12.1
If the total impulse developed by 750 g fuel is 1200 N s, determine the specific impulse of the fuel in seconds.

Solution
Given: impulse $I = 1200$ N s and the mass of the fuel is $m = 0.75$ kg.
 Therefore, the weight of the fuel is

$$w = mg$$
$$= 0.75 \times 9.81$$
$$= 7.36 \text{ N}$$

Thus, the specific impulse of the fuel per unit weight is

$$I_w = \frac{I}{w}$$
$$= \frac{1200}{7.36}$$
$$= \boxed{163 \text{ seconds}}$$

12.6 Rocket Equation

In defining the performance of a rocket, the so-called rocket equation is used. This is derived by noting that if M is the mass of the rocket vehicle at any instant of time and V is its velocity at this time, as shown in Figure 12.6.

Force acting on rocket = mass of rocket × acceleration of rocket, if the rocket is moving in a vertical direction. Thus,

$$T = M\frac{dV}{dt} + Mg \qquad (12.10)$$

Using Eq. (12.5), this can be expressed as

$$M\frac{dV}{dt} = -\frac{dm_f}{dt}V_e - Mg \qquad (12.11)$$

But

$$M = m_s + m_f$$

where m_s is the mass of the vehicle including motor structure and payload, that is, the dry mass, which does not change with time, and m_f is the fuel mass. Thus,

$$\frac{dM}{dt} = \frac{dm_f}{dt} \qquad (12.12)$$

Hence, Eq. (12.11) can be written as

$$M\frac{dV}{dt} = -\frac{dM}{dt}V_e - Mg \qquad (12.13)$$

Figure 12.6 A moving rocket.

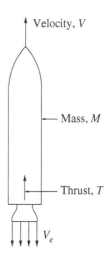

If the exhaust velocity can be assumed constant, this equation can be integrated to give

$$-V_e \int_{M_i}^{M_f} \frac{dM}{M} = \int_{V_i}^{V_f} dV + g \tag{12.14}$$

where M_i and M_f are the initial and final masses of the rocket vehicle and V_i and V_f are its initial and final velocities. Integrating Eq. (12.14), we get

$$V_e \ln \left(\frac{M_i}{M_f} \right) = \left(V_f - V_i \right) + gt \tag{12.15}$$

Defining

$$\Delta V = \left(V_f - V_i \right) \tag{12.16}$$

we have

$$\Delta V = V_e \ln \left(\frac{M_i}{M_f} \right) - gt \tag{12.17}$$

In the above derivation, the gravitational acceleration g has been assumed to be constant. However, this may not be a justifiable assumption always.

If the vehicle initially has a velocity of zero, the velocity achieved when all the fuel has been used, this velocity being called the burnout velocity, V_b, is given by Eq. (12.17) as

$$\boxed{V_b = V_e \ln \left(\frac{M_i}{M_f} \right) - gt} \tag{12.18}$$

This is called the *rocket equation*. The quantity M_i/M_f is called the *rocket mass ratio*. Equation may also be expressed as

$$V_b = V_e \ln \left(\frac{m_s + m_f}{m_s} \right) - gt \tag{12.19}$$

A high exhaust velocity has historically been a driving force for rocket design: payload fractions depend strongly upon the exhaust velocity, as seen from Eq. (12.17). Variation of exhaust velocity with payload fraction is shown in Figure 12.7.

Note that in the derivation of rocket velocity, the effects of atmospheric drag have been neglected. If this is accounted for, Eq. (12.17), for example, will become

$$\Delta V = V_e \ln \left(\frac{M_i}{M_f} \right) - gt - \int_0^t \frac{D}{M} dt$$

where D is the drag force acting on the rocket at any instant of time. Its value depends on the size and shape of the vehicle, the velocity, the Mach number, and the local properties of the atmosphere through which the vehicle is passing.

12.7 Efficiency

When the total kinetic energy of the rocket and its exhaust are taken into consideration, the highest propulsive efficiency, η_p, occurs when the exhaust velocity is equal to the instantaneous rocket velocity, as shown in Figure 12.8. The overall propulsive efficiency η is the efficiency with which the energy contained in a vehicle's propellant is converted into useful energy to replace losses due to air drag, gravity, and acceleration.

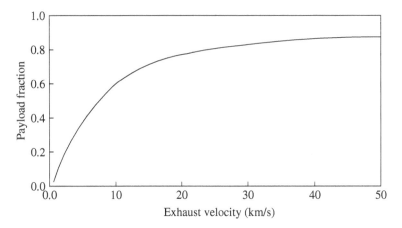

Figure 12.7 Exhaust velocity variation with payload fraction.

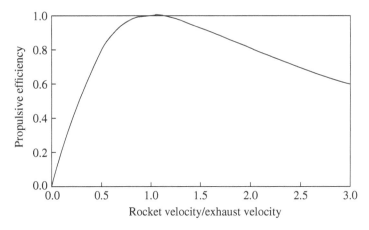

Figure 12.8 Propulsive efficiency variation with V/V_e.

Mathematically, it is represented as

$$\eta = \eta_c \eta_p$$

where η_c is the cycle efficiency and η_p is the propulsive efficiency. The cycle efficiency, in percent, is the proportion of energy that can be derived from the energy source that is converted to mechanical energy by the engine.

The cycle efficiency, η_c, of a rocket engine is usually high due to the high combustion temperatures and pressures and long nozzle. The value varies slightly with altitude due to the atmospheric pressure on the outside of the nozzle/engine but can be up to 70%. Most of the remainder is lost as heat energy in the exhaust.

Rocket engines have a slightly different propulsive efficiency (η_p) than air-breathing jet engines as the lack of intake air changes the form of the equation. This also means that rockets are able to exceed their exhaust velocity:

$$\eta_p = \frac{2\,(V/V_e)}{1 + (V/V_e)^2}$$

12.8 Trajectories

Spacecraft today essentially travel by being given an impulse that places them on a trajectory in which they coast from one point to another, perhaps with other impulses or gravity assists along the way. The gravity fields of the Sun and planets govern such trajectories. Rockets launched through atmospheres face additional complications, such as air friction and winds. Most of the present discussion treats this type of trajectory.

Advanced propulsion systems that efficiently travel throughout the Solar System will be required for human exploration, settlement, and accessing space resources. Rather than coasting, advanced systems will thrust for most of a trip, with higher exhaust velocities but lower thrust levels. These complicated trajectories require advanced techniques for finding optimum solutions.

12.8.1 Newton's Laws of Motion

The fundamental laws of mechanical motion were first formulated by Sir Isaac Newton (1643–1727) and were published in his *Philosophia Naturalis Principia Mathematica*. They are the following:

1. Everybody continues in its state of rest or of uniform motion in a straight line except insofar as it is compelled to change that state by an external impressed force.
2. The rate of change of momentum of the body is proportional to the impressed force and takes place in the direction in which the force acts.
3. To every action there is an equal and opposite reaction ($dp/dt = F$).
4. Calculus, invented independently by Newton and Gottfried Leibniz (1646–1716), plus Newton's laws of motion are the mathematical tools needed to understand rocket motion.

12.8.2 Newton's Laws of Gravitation

To calculate the trajectories for planets, satellites, and space probes, the additional relation required is Newton's law of gravitation.

Every particle of matter attracts every other particle of matter with a force directly proportional to the product of their masses and inversely proportional to the square of the distance between them.

Symbolically, the force is

$$F = -G\, m_1\, m_2\, e_r / r^2$$

where $G = 6.67 \times 10^{-11}$ m^3/(s^2 kg), m_1 and m_2 are the interacting masses (kg), r is the distance between them (m), and e_r is a unit vector pointing between them.

12.8.3 Kepler's Laws of Planetary Motion

The discovery of the laws of planetary motion owed a great deal to Tycho Brahe's (1546–1601) observations, from which Johannes Kepler (1571–1630) concluded that the planets move in elliptical orbits around the Sun. First, however, Kepler spent many years trying to fit the orbits of the five then-known planets into a framework based on the five regular platonic solids. The laws are the following:

1. The planets move in ellipses with the sun at one focus. That is, the path of the planets about the sun is elliptical in shape, with the center of the sun being located at one focus (The Law of Ellipses).

2. Areas swept out by the radius vector from the sun to a planet in equal times are equal. In other words, an imaginary line drawn from the centre of the sun to the centre of the planet will sweep out equal areas in equal intervals of time (The Law of Equal Areas).

3. The square of the period of revolution is proportional to the cube of the semi-major axis. That is, $T^2 = \text{constant} \times a^3$. The ratio of the squares of the periods of any two planets is equal to the ratio of the cubes of their average distances from the sun (The Law of Harmonies).

In a central-force gravitational potential, bodies will follow conic sections:

$$r = a_0/(1 + e \, \cos \, \theta)$$
$$a = a_0/(1 - e^2)$$

where e is eccentricity and a_0 is the semi-major axis.

Special cases:

$$e > 1 \quad E > 0 \text{ hyperbola}$$
$$e = 1 \quad E = 0 \text{ parabola}$$
$$e < 1 \quad E < 0 \text{ ellipse}$$
$$e = 0 \quad E < 0 \text{ circle}$$

where E is specific mechanical energy.

12.8.4 Some Important Equations of Orbital Dynamics

The relations for circular velocity, escape velocity, and energy of a vehicle following a conic section are important equations of orbital dynamics. They are listed below:

Circular velocity:

$$v_{\text{cir}} = \left(\frac{GM}{R} \right)^{1/2}$$

Escape velocity:

$$v_{\text{esc}} = \left(\frac{2GM}{R} \right)^{1/2}$$

Energy of a vehicle following a conic section:

$$E_{\text{conic}} = \frac{-GMm}{2a}$$

where a is the semi-major axis.

12.8.5 Lagrange Points

The Lagrange (sometimes called Liberation) points are positions of equilibrium for a body in a two-body system. The points L_1, L_2, and L_3 lie on a straight line through the other two bodies and are points of unstable equilibrium. That is, a small perturbation will cause the third body to drift away. The L_4 and L_5 points are at the third vertex of an equilateral triangle formed with the other two bodies; they are points of stable equilibrium. The approximate positions for the Earth–Moon or Sun–Earth Lagrange points are shown in Figure 12.9.

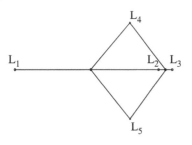

Figure 12.9 Approximate positions for the Earth–Moon or Sun–Earth Lagrange points.

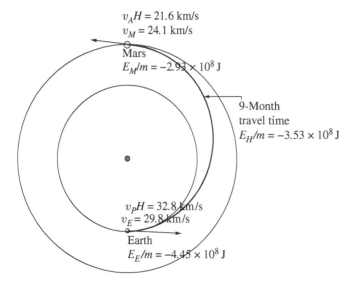

Figure 12.10 Energy per unit mass on the circular orbit and Hohmann trajectory.

12.8.6 Hohmann Minimum-Energy Trajectory

The minimum-energy transfer between circular orbits is an elliptical trajectory called the Hohmann trajectory. It is shown for the Earth–Mars case, where the minimum total change in velocity (δV) expended is 5.6 km/s. The values of the energy per unit mass on the circular orbit and Hohmann trajectory are shown in Figure 12.10, along with the velocities at perihelion (closest to Sun) and aphelion (farthest from Sun) on the Hohmann trajectory and the circular velocity in Earth or Mars orbit. The differences between these velocities are the required δV values in the rocket equation.

12.8.7 Gravity Assist

Gravity assists enable or facilitate many missions. A spacecraft arrives within the sphere of influence of a body with a so-called hyperbolic excess velocity equal to the vector sum of its incoming velocity and the planet's velocity. In the planet's frame of reference, the direction of the spacecraft's velocity changes, but not its magnitude. In the spacecraft's frame of reference, the net result of this trade-off of momentum is a small change in the planet's velocity and a very large δV for the spacecraft. Starting from an Earth–Jupiter Hohmann trajectory and performing a Jupiter flyby at one Jovian radius, as shown in Figure 12.11, the hyperbolic excess velocity V_h is approximately 5.6 km/s, and the angular change in direction is about 160°.

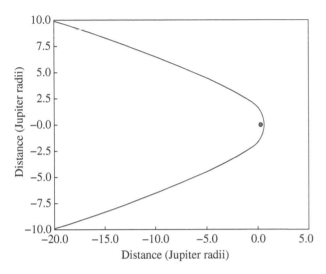

Figure 12.11 Jupiter flight path at one Jovian radius, starting from an Earth–Jupiter Hohmann trajectory.

12.9 High-Exhaust-Velocity, Low-Thrust Trajectories

The simplest high-exhaust-velocity analysis splits rocket masses into three categories:

1. Power plant and thruster system mass, M_w.
2. Payload mass, M_l (note that this includes all structure and other rocket mass that would be treated separately in a more sophisticated definition).
3. Propellant mass, M_p.
   Mission power-on time τ

Total mass:

$$M_0 = M_w + M_l + M_p$$

Empty mass:

$$M_e = M_w + M_l$$

Specific power, α (kW/kg):

$$\alpha \ (\text{kW/kg}) \equiv \frac{P_w}{M_w} \equiv \frac{P_{\text{thrust}}}{M_{\text{power system}} + M_{\text{thrust system}}}$$

Propellant flow rate:

$$\dot{M} = \frac{M_p}{\tau}$$

Thrust power:

$$P_w = \frac{1}{2}\dot{M}V_j^2 = \frac{M_p V_j^2}{2\tau}$$

Thrust:

$$F = \dot{M}V_j = \frac{M_p V_j}{\tau}$$

12.9.1 High-Exhaust-Velocity Rocket Equation

Assume constant exhaust velocity, V_{ex}, which greatly simplifies the analysis. The empty (final) mass in the Tsiolkovsky rocket equation now becomes $M_w + M_l$, so

$$\frac{M_f}{M_0} = \frac{M_a}{M_0} = \frac{M_w + M_l}{M_w + M_l + M_p} = \exp\left(\frac{-u}{V_j}\right)$$

where u measures the energy expended in a manner analogous to δv. After some messy but straightforward algebra, we get the high-exhaust-velocity rocket equation

$$\frac{M_L}{M_0} = \exp\left(\frac{-u}{V_j}\right) - \frac{V_j^2}{2\alpha\tau}\left[1 - \exp\left(\frac{-u}{V_j}\right)\right]$$

Note that a chemical rocket effectively has $M_w = 0$, implying that $\alpha = \infty$, and the Tsiolkovsky equation ensues. The quantity $\alpha \times \tau$ is the energy produced by the power and thrust system during a mission with power-on time τ divided by the mass of the propulsion system. It is called the specific energy of the power and thrust system.

Relating the specific energy to a velocity through $E = \frac{1}{2}mv^2$ gives the definition of a very important quantity, the characteristic velocity:

$$v_{ch} \equiv (2\alpha\tau)^{1/2}$$

The payload fraction for a high-exhaust-velocity rocket becomes

$$\frac{M_L}{M_0} = \exp\left(\frac{-u}{V_j}\right) - \frac{V_j^2}{v_{ch}^2}\left[1 - \exp\left(\frac{-u}{V_j}\right)\right]$$

The variation of $\frac{M_L}{M_0}$ with $\frac{V_j}{v_{ch}}$ for different values of $\frac{u}{v_{ch}}$ is shown in Figure 12.12.

Analysing a trajectory using the characteristic velocity method requires an initial guess for τ plus some iterations. The minimum energy expended will always be more than the Hohmann-trajectory energy. The payload capacity of a fixed-velocity rocket vanishes at $u = 0.81v_{ch}$, where $V_j = 0.5v_{ch}$.

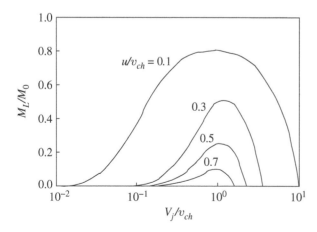

Figure 12.12 Payload fraction variation with $\dfrac{V_j}{v_{ch}}$.

Substituting these values into the rocket equation gives

$$\frac{M_0}{M_e} = \exp\left(\frac{-u}{V_j}\right) = \exp\left(\frac{0.81}{0.5}\right) \approx 5$$

12.10 Plasma and Electric Propulsion

A plasma propulsion engine is a type of *electric propulsion* that generates thrust from a quasi-neutral *plasma*. This is in contrast to ion beam engines, which generates thrust through extracting an ion current from plasma source, which is then accelerated to high velocities using grids/anodes. These exist in many forms (see electric propulsion). Plasma thrusters do not typically use high-voltage grids or anodes/cathodes to accelerate the charged particles in the plasma, but rather uses currents and potentials that are generated internally in the plasma to accelerate the plasma ions. While this results in a lower exhaust velocities by virtue of the lack of high accelerating voltages, this type of thruster has a number of interesting advantages. The lack of high-voltage grids of anodes removes a possible limiting element as a result of grid ion erosion. The plasma exhaust is 'quasi-neutral', which means that ion and electrons exist in equal number, which allows simply ion–electron recombination in the exhaust to neutralise the exhaust plume, removing the need for an electron gun (hollow cathode). This type of thruster often generates the source plasma using radio frequency of microwave energy, using an external antenna. This fact, combined with the absence of hollow cathodes (which are very sensitive to all but the few noble gases), allows the intriguing possibility of being able to use this type of thruster on a huge range of propellants, from argon to carbon dioxide to air mixtures to astronaut urine. Plasma engines are better suited for long-distance *interplanetary space travel* missions.

In recent years, many agencies have developed several forms of plasma-fuelled engines, including the European Space Agency, Iranian Space Agency, and Australian National University, which have co-developed a more advanced type described as a double-layer thruster. However, this form of plasma engine is only one of many types.

12.10.1 Types of Plasma Engines

The following are the types of plasma engines developed.

Helicon double-layer thruster: A helicon double-layer thruster uses radio waves to create a plasma and a magnetic nozzle to focus and accelerate the plasma away from the rocket engine. A mini-helicon plasma thruster, ideal for space manoeuvres, runs off of nitrogen, and the fuel has an exhaust velocity (specific impulse) 10 times that of chemical rockets.

Magnetoplasmadynamic thrusters: Magnetoplasmadynamic thrusters (MPD) uses the Lorentz force (a force resulting from the interaction between a magnetic field and an electric current) to generate thrust – the electric charge flowing through the plasma in the presence of a magnetic field causing the plasma to accelerate due to the generated magnetic force.

Hall effect thrusters: Hall effect thrusters combine a strong localised static magnetic field perpendicular to the electric field created between an upstream anode and a downstream cathode called neutraliser to create a 'virtual cathode' (area of high electron density) at the exit of the device. This virtual cathode then attracts the ions formed inside the thruster closer to the anode. Finally, the accelerated ion beam is neutralised by some of the electrons emitted by the neutraliser.

Electrodeless plasma thrusters: Electrodeless plasma thrusters use the ponderomotive force[1] that acts on any plasma or charged particle when under the influence of a strong electromagnetic energy gradient to accelerate the plasma.

SPT series: Serial production started in Soviet Union in 1970s. One of the early variants, SPT-100, is now produced under license by European Snecma Moteurs under the name PPS-1350. SPT-290 has a thrust of 1.5 N, power of 5–30 kW specific impulse of 30 km/s, efficiency of 65%, and weight of 23 kg.

VASIMR: Variable Specific Impulse Magnetoplasma Rocket or VASIMR works by using radio waves to ionise a propellant into a plasma and then a magnetic field to accelerate the plasma out of the back of the rocket engine to generate thrust. The VASIMR is currently being developed by the private company Ad Astra Rocket Company, headquartered in Houston, TX, with the help of an NS Canada-based company Nautel, producing the 200 kW RF generators for ionising propellant. Some of the components and 'Plasma Shoots' experiments are tested in a laboratory settled in Liberia, Costa Rica. This project is led by former NASA astronaut Dr. Franklin Chang-Daz (CRC-USA). Recently the Costa Rican Aerospace Alliance announced the cooperation to this project by developing an exterior support device for the VASIMR to be fitted in the exterior of the International Space Station, as part of the plan to test the VASIMR in space. The VF-200 engine could reduce the duration of flight from the Earth to Jupiter or Saturn from 6 years to 14 months and Mars from 4 months to 39 days.

12.11 Pulsed Plasma Thruster

A pulsed plasma thruster (PPT), also known as a plasma jet engine, is a form of electric spacecraft propulsion. PPTs are generally considered the simplest form of electric spacecraft propulsion and were the first form of electric propulsion to be flown in space, having flown on two Soviet probes (Zone 2 and Zone 3) starting in 1964. PPTs are generally flown on spacecraft with a surplus of electricity from abundantly available solar energy.

12.11.1 Operating Principle

Most PPTs use a solid material (normally polytetrafluoroethylene PTFE, more commonly known as Teflon, which is a synthetic fluoropolymer of tetrafluoroethylene that has numerous applications) for propellant, although a minority use liquid or gaseous propellants. The first stage in PPT operation involves an arc of electricity passing through the fuel, causing ablation and sublimation of the fuel. The heat generated by this arc causes the resultant gas to turn into plasma, thereby creating a charged gas cloud. Due to the force of the ablation, the plasma is propelled at low speed between two charged plates (an anode and cathode). Since the plasma is charged, the fuel effectively completes the circuit between the two plates, allowing a current to flow through the plasma. This flow of electrons generates a strong electromagnetic field that then exerts a Lorentz force on the plasma, accelerating the plasma out of the PPT exhaust at high velocity. The pulsing occurs due to the time needed to recharge the plates following each burst of fuel and the time between each arc. The frequency of pulsing is normally very high, and so it generates an almost continuous and smooth thrust. While the thrust is very low, a PPT can operate continuously for extended periods of time, yielding a large final speed.

---

1 A nonlinear force that a charged particle experiences in an inhomogeneous oscillating electromagnetic field.

The energy used in each pulse is stored in a capacitor. By varying the time between each capacitor discharge, the thrust and power drawn out of the PPT can be varied allowing versatile use of the system.

The equation for the change in velocity of a spacecraft is given by the rocket equation as follows:

$$\Delta v = v_e \ln \frac{m_0}{m_1}$$

where Δv is the maximum change of speed of the vehicle (with no external forces acting), v_e is the effective exhaust velocity ($v_e = I_{sp} \cdot g_0$ where I_{sp} is the specific impulse expressed as a time period and g_0 is standard gravity), m_0 is the initial total mass including propellant, and m_1 is the final total mass.

PPTs have much higher exhaust velocities than chemical propulsion engines but have a much smaller fuel flow rate. From the Tsiolkovsky equation stated above, this results in a proportionally higher final velocity of the propelled craft. The exhaust velocity of a PPT is of the order of tens of km/s, while conventional chemical propulsion generates thermal velocities in the range of 2–4.5 km/s. Due to this lower thermal velocity, chemical propulsion units become exponentially less effective at higher vehicle velocities, necessitating the use of electric spacecraft propulsion such as PPTs. It is therefore advantageous to use an electric propulsion system such as a PPT to generate high interplanetary speeds in the range of 20–70 km/s.

The advantages of PPTs are that they are robust due to their inherently simple design (relative to other electric spacecraft propulsion techniques) and draw very little electrical power relative to other comparable thrusters. As an electric propulsion system, PPTs benefit from reduced fuel consumption compared with traditional chemical rockets, reducing launch mass and therefore launch costs, as well as high specific impulse improving performance.

However, due to energy losses caused by late time ablation and rapid conductive heat transfer from the propellant to the rest of the spacecraft, propellant efficiency is very low compared with other forms of electric propulsion, at around just 10%.

PPTs are well suited for relatively small spacecraft with a mass of less than 100 kg (particularly CubeSats) for roles such as attitude control, station keeping, de-orbiting manoeuvres, and deep space exploration. Using PPTs could double the life span of these small satellite missions without significantly increasing complexity or cost due to the inherent simplicity and relatively low cost nature of PPTs. A PPT was flown by NASA in November 2000, as a flight experiment on the Earth Observing-1 spacecraft. The thrusters successfully demonstrated the ability to perform roll control on the spacecraft and also demonstrated that the electromagnetic interference from the pulsed plasma did not affect other spacecraft systems. PPTs are also an avenue of research used by universities for starting experiments with electric propulsion due to the relative simplicity and lower costs involved with PPTs as opposed to other forms of electric propulsion such as Hall effect ion thrusters.

Example 12.2

A rocket flies at an altitude of 1000 m. If the combustion chamber pressure and temperature are 10 atm and 1200 K and the nozzle is adapted, determine the mass flow rate through the nozzle, exit velocity, and the thrust generated, if the throat area is 0.3 m². Assume the specific heat ratio of the combustion products is 1.22 and the gas constant is 275 m²/(s² K).

Solution

Given: $p_0 = 10\,\text{atm}$, $T_0 = 1200\,\text{K}$, $\gamma = 1.22$, $A_{th} = 0.3\,\text{m}^2$, $p_e = p_a$, and $R = 275\,\text{m}^2/(\text{s}^2\,\text{K})$.

At 1000 m altitude, from atmospheric table, the pressure is $p_a = 89.9\,\text{kPa}$.

By isentropic equation,

$$\frac{p_0}{p_e} = \left(1 + \frac{\gamma - 1}{2} M_e^2\right)^{\gamma/(\gamma-1)}$$

$$\frac{10 \times 101.325}{89.9} = \left(1 + \frac{1.22 - 1}{2} M_e^2\right)^{\gamma/(\gamma-1)}$$

$$11.27 = \left(1 + 0.11\, M_e^2\right)^{1.22/(1.22-1)}$$

$$11.27 = \left(1 + 0.11\, M_e^2\right)^{5.545}$$

$$1 + 0.11\, M_e^2 = 11.27^{1/(5.545)}$$

$$= 1.548$$

$$M_e = \sqrt{\frac{0.548}{0.11}}$$

$$= 2.23$$

The stagnation density is

$$\rho_0 = \frac{p_0}{RT_0}$$

$$= \frac{10 \times 101\,325}{275 \times 1200}$$

$$= 3.07 \text{ kg/m}^3$$

By isentropic relation,

$$\frac{\rho_0}{\rho^*} = \left(1 + \frac{\gamma - 1}{2} M^{*2}\right)^{1/(\gamma-1)}$$

$$= \left(\frac{\gamma + 1}{2}\right)^{1/(1.22-1)}$$

$$= 1.11^{4.545}$$

$$= 1.6$$

Note that for choked throat the local Mach number $M^* = 1$. Thus, the density at the throat is

$$\rho^* = \frac{\rho_0}{1.6}$$

$$= \frac{3.07}{1.6}$$

$$= 1.92 \text{ kg/m}^3$$

The temperature at the throat is

$$T^* = \frac{T_0}{\left(1 + \frac{\gamma-1}{2} M^{*2}\right)}$$

$$= \frac{1200}{1.11}$$

$$= 1081 \text{ K}$$

The speed of sound at the throat is

$$a^* = \sqrt{\gamma R T^*}$$

$$= \sqrt{1.22 \times 275 \times 1081}$$

$$= 602.22 \text{ m/s}$$

Therefore, the mass flow rate becomes

$$\dot{m} = \rho^* A^* V^*$$
$$= \rho^* A^* a^*$$
$$= 1.92 \times 0.3 \times 602.22$$
$$= \boxed{346.88 \text{ kg/s}}$$

The exit temperature is

$$\frac{T_0}{T_e} = 1 + \frac{\gamma - 1}{2} M_e^2$$
$$= 1 + \frac{1.22 - 1}{2} \times 2.23^2$$
$$= 1.547$$
$$T_e = \frac{T_0}{1.547}$$
$$= \frac{1200}{1.547}$$
$$= 775.69 \text{ K}$$

The exit velocity is

$$V_e = M_e a_e$$
$$= M_e \times \sqrt{\gamma R T_e}$$
$$= 2.23 \times \sqrt{1.22 \times 275 \times 775.69}$$
$$= 2.23 \times 510.14$$
$$= \boxed{1137.61 \text{ m/s}}$$

The thrust generated is

$$Th = \dot{m} V_e$$
$$= 346.88 \times 1137.61$$
$$= 394\ 614 \text{ N}$$
$$= \boxed{394.614 \text{ kN}}$$

The mass flow rate can also be expressed as

$$\dot{m} = \sqrt{\frac{\gamma}{R T_0} \left(\frac{2}{\gamma + 1}\right)^{(\gamma+1)/(\gamma-1)}} p_0 A^*$$

Therefore,

$$\dot{m} = \sqrt{\left(\frac{1.22}{275 \times 1200}\right) \times \left(\frac{2}{1.22 + 1}\right)^{(2.22)/(0.22)}} p_0 A^*$$
$$= (1.136 \times 10^{-3}) \times (10 \times 101\ 325) \times 0.3$$
$$= 345.32 \text{ kg/s}$$

12.12 Summary

A rocket engine is an engine that produces a force (a thrust) by creating a high-velocity output without using any of the constituents of the 'atmosphere' in which the rocket is operating. It can operate in any part of the atmosphere and outside the atmosphere that makes it ideal for space propulsion.

In a chemical rocket, a fuel and an oxidiser are usually supplied to the combustion chamber of the rocket. There are two types of chemical rockets, *liquid-propellant rockets* and *solid-propellant rockets*.

In a non-chemical rocket, the high efflux velocity from the rocket is generated without any chemical reaction taking place. For example, a gas could be heated to high pressure and temperature by passing it through a nuclear reactor, and it could then be expanded through a nozzle to give a high efflux velocity.

The issues associated with the engine design are the following:

1. Heat transfer: This involves *radiative cooling*, radiating heat to space or conducting it to the atmosphere; *regenerative cooling*, running cold propellant through the engine before exhausting it; *boundary-layer cooling*, aiming some cool propellant at the combustion chamber walls; and transpiration cooling, diffusing coolant through porous walls.
2. Nozzles: *Rocket nozzles* are usually of an expansion–deflection design.

The International Space Station is the centrepiece of our human spaceflight activities in low-Earth orbit.

The thrust is equal in magnitude to the force acting on the rocket:

$$T = \dot{m}V_e + (p_e - p_a)A_e$$

Specific impulse, I_{sp}, is the thrust per unit mass flow rate of propellant. The specific impulse is mainly dependent on the type of propellant used.

In defining the performance of a rocket, rocket equation is used. It can be expressed as

$$\boxed{V_b = V_e \ln\left(\frac{M_i}{M_f}\right) - gt}$$

The quantity M_i/M_f is called the *rocket mass ratio*.

The highest propulsive efficiency, η_p, occurs when the exhaust velocity is equal to the instantaneous rocket velocity. The overall propulsive efficiency η is the efficiency with which the energy contained in a vehicle's propellant is converted into useful energy to replace losses due to air drag, gravity, and acceleration.

The cycle efficiency, η_c, of a rocket engine is usually high due to the high combustion temperatures and pressures and long nozzle employed.

Spacecraft today essentially travel by being given an impulse that places them on a trajectory in which they coast from one point to another, perhaps with other impulses or gravity assists along the way. The gravity fields of the Sun and planets govern such trajectories.

To calculate the trajectories for planets, satellites, and space probes, the additional relation required is Newton's law of gravitation.

The planets move in elliptical orbits around the Sun. The laws are the following. (i) The planets move in ellipses with the sun at one focus. (ii) Areas swept out by the radius vector from the sun to a planet in equal times are equal. (iii) The square of the period of revolution is proportional to the cube of the semi-major axis.

In a central-force gravitational potential, bodies will follow conic sections.

The relations for circular velocity, escape velocity, and energy of a vehicle following a conic section are important equations of orbital dynamics.

The Lagrange (sometimes called Liberation) points are positions of equilibrium for a body in a two-body system.

The minimum-energy transfer between circular orbits is an elliptical trajectory called the Hohmann trajectory.

A plasma propulsion engine is a type of *electric propulsion* that generates thrust from a quasi-neutral *plasma*.

The following are the types of plasma engines developed: (i) helicon double-layer thruster, (ii) MPD, (iii) Hall effect thrusters, and (iv) electrodeless plasma thrusters.

A PPT, also known as a plasma jet engine, is a form of electric spacecraft propulsion. The advantages of PPTs are that they are robust due to their inherently simple design (relative to other electric spacecraft propulsion techniques) and draw very little electrical power relative to other comparable thrusters.

Problems

12.1 A spacecraft's engine ejects mass at a rate of 26 kg/s with an exhaust velocity of 3000 m/s. The pressure at the nozzle exit is 5.6 kPa, and the exit area is 0.53 m^2. What is the thrust of the engine in a vacuum?

12.2 A spacecraft's dry mass is 70 000 kg, and the effective exhaust gas velocity of its main engine is 3000 m/s. How much propellant must be carried if the propulsion system is to produce a total ΔV of 680 m/s?

References

1 Darrel, D.M. (2014). *Illustrated History of Wind Power Development*. TelosNet Home Page: Littleton, CO. http://ww.telosnet.com/wind/early.html.

2 Berliner, D. (1997). *Aviation: Reaching for the Sky*, 128. Oliver Press, Inc.. ISBN 1-881508-33-1.

3 Ovid, G.H. (2001). *The Metamorphoses*. Signet Classics. ISBN 0-451-52793-3.

4 Newton, I. (1726). *Philosophiae Naturalis Principia Mathematica*, London, 1687; Cambridge, 1713; London, 1726: Mathematical Principles of Natural Philosophy.

5 von Karman, T. (2004). *Aerodynamics: Selected Topics in the Light of Their Historical Development*. Dover Publications. ISBN 0-486-43485-0.

6 The Editors of Encyclopaedia Britannica (1988). *Hydrodynamica*. Britannica Online Encyclopedia.

7 Cayley, G. (1773–1857). *U.S Centennial of Flight Commission - Sir George Cayley*. U.S Centennial of Flight Commission.

8 Cayley, G. (1809–1810). "On aerial navigation" Part 1–3. *Nicholson's Journal of Natural Philosophy*.

9 d'Alembert, J. (1752). *Essai d'une nouvelle theorie de la resistance des fluides*. Bruxelles, Culture et Civilisation, 1966.

10 Kirchhoff, G. (1869). Zur Theorie freier Flussigkeitsstrahlen. *Journal fur die reine und angewandte Mathematik* 70: 289–298.

11 Rayleigh, L. (1876). On the resistance of fluids. *Philosophical Magazine* 5 (2): 430–441.

12 Navier, C.L.M.H. (1827). Memoire sur les lois du mouvement des fluides. *Memoires de l'Academie des Sciences* 6: 389–440.

13 Stokes, G.G. (1845). On the theories of the internal friction of fluids in motion. *Transaction of the Cambridge Philosophical Society* 8: 287–319.

14 Reynolds, O. (1883). An experimental investigation of the circumstances which determine whether the motion of water shall be direct or sinuous and of the law of resistance in parallel channels. *Philosophical Transactions of the Royal Society of London A* 174: 935–982.

15 Renard, C. (1889). Nouvelles experiences sur la resistance de l'air. *L'Aeronaute* 22: 73–81.

16 Chanute, O. (1997). *Progress in Flying Machines*. Dover Publications. ISBN 0-486-29981-3.

17 Lanchester, F.W. (1907). *Aerodynamics*. London: Archibald Constable & Co. Ltd.

18 Busemann, A. (1960). Ludwig Prandtl. 1875–1953. Biogr. Mems of Fell. R. l Soc. 5: 193–205. doi:10.1098/rsbm.1960.0015.

19 Bodie, W.M. (2001). *The Lockheed P-38 Lightning*, 174–175. Widewing Pub.

20 Ackeret, J. (1925). Luftkrafte auf Flugel, die mit der grosserer als Schallgeschwindigkeit bewegt werden. *Zeitschrift fur Flugtechnik und Motorluftschiffahrt* 16: 72–74.

21 Anderson, J.D. (2007). *Fundamentals of Aerodynamics*, 4e. McGraw-Hill. ISBN 0-07-125408-0.

Introduction to Aerospace Engineering: Basic Principles of Flight, First Edition. Ethirajan Rathakrishnan.
© 2021 John Wiley & Sons, Inc. Published 2021 by John Wiley & Sons, Inc.
Companion Website: www.wiley.com/go/Rathakrishnan/IntroductiontoAerospaceEngineering

22 Rathakrishnan, E. (2012). *Fluid Mechanics - An Introduction*, 3e. Delhi: PHI Learning.

23 Katz, J. (1991). *Low-Speed Aerodynamics: From Wing Theory to Panel Methods*, McGraw-Hill Series in Aeronautical and Aerospace Engineering. New York: McGraw-Hill. ISBN 0-07-050446-6.

24 Rathakrishnan, E. (2020). *Gas Dynamics*, 7e. Delhi: PHI Learning.

25 ISO 2533:1975 (1975). *Standard atmosphere*. International Organization for Standardization.

26 Gyatt, G. (2006-01-14). The Standard Atmosphere. A mathematical model of the 1976 U.S. Standard Atmosphere.

27 Auld, D.J. and Srinivas, K. (2008). Properties of the atmosphere. http://www.aeromech.usyd.edu.au/aero/atmosphere/Retrieved2008-03-13.

28 Batchelor, G.K. (1967). *An Introduction to Fluid Dynamics*. Cambridge University Press.

29 International Civil Aviation Organisation (1993). *Manual of the ICAO Standard Atmosphere (extended to 80 kilometers (262 500 feet))*, Doc 7488-CD, 3rd Edition. ISBN 92-9194-004-6.

30 Secretary General of the International Civil Aviation Organization (1958). *U.S. Extension to the ICAO Standard Atmosphere*. Washington, DC: U.S. Government Printing Office.

31 U.S. Standard Atmosphere (1962). National Aeronautics and Space Administration, United States Air Force, United States Weather Bureau, Washington, DC.

32 National Oceanic and Atmospheric Administration; National Aeronautics and Space Administration United States Air Force (1966). *U.S. Standard Atmosphere Supplements*. Washington, DC: U.S. Government Printing Office.

33 U.S. Standard Atmosphere (1976). National Oceanic and Atmospheric Administration, National Aeronautics and Apace Administration, United States Air Force, Washington, D.C.

34 NASA (1976). National Aeronautics and Space administration (NASA) U.S. Standard Atmosphere, Report/Patent Number *NOAA-S/T-76-1562/ NASA-TM-X-74335*, Washington, DC.

35 Tomasi, C., Vitake, V., and De Santis, L.V. (1998). Relative optical mass functions for air, water vapour, ozone and nitrogen dioxide in atmospheric models presenting different latitudinal and seasonal conditions. *Meteorology and Atmospheric Physics* 65 (1): 11–30. Bibcode 1998MAP....65...11T. https://doi.org/10.1007/BF01030266. Retrieved 2007-12-31. "The ISO (International Organization for Standardization) Standard Atmosphere, 1972. This model is identical to the present Standard Atmospheres of ICAO (International Civil Aviation Organization) and WMO (World Meteorological Organization) up to a height of 32 km".

36 Rathakrishnan, E. (2013). *Theoretical Aerodynamics*. New Jersey: Wiley.

Appendix A

Table A.1 SI units and their conversion to US units.

	SI	US
Length	Metre (m)	Foot (ft) = 0.3048 m
		Yard (yd) = 0.9144 m
	Centimetre (cm) = 10^{-2} m	Inch (in.) = 25.4 mm
	Kilometre (km) = 10^{3} m	Mile (mi) = 1.609 km
Area	Square metre (m^2)	Square foot (ft^2) = 0.0929 m^2
		square yard (yd^2) = 0.8361 m^2
	Square centimetre (cm^2) = 10^{-4} m^2	Square inch ($in.^2$) = 6.452 cm^2
	Hectare (ha) = 10^4 m^2	acre (A or ac) = 43 560 ft^2 = 0.4047 ha.
	Square kilometre (km^2) = 10^6 m^2 = 100 ha	Sqaure mile (mi^2) = 640 A = 2.590 km^2
Volume	Litre (l or L)	gallon (gal) = 3.785 Litre or l
		Cubic foot ft^3 = 7.481 gal = 28.31 l
	Cubic centimetre (cm^3) = 1 millilitre (ml)	cubic inch ($in.^3$ = 16.39 cm^3)
	cubic metre m^3 = 10^3 l	cubic yard (yd^3) = 0.7645 m^3
Temperature	Degree Celsius (°C)	Degree Fahrenheit (°F) = 9/5(°C) + 32
Mass	Kilogram (kg)	Pound (lb) = 0.4536 kg
	gram (g) = 10^{-3} kg	ounce (oz) = 1/16 lb = 28.35 g
	Metric tonne or metric ton (t) = 10^3 kg	Short ton (commonly caller "ton") =
		2000 lb = 0.9071 t
		Long ton = 2240 lb = 1.016 t
	quintal (q) = 10^2 kg	Hundredweight (cwt) = 112 lb = 5080.23 g

(Continued)

Introduction to Aerospace Engineering: Basic Principles of Flight, First Edition. Ethirajan Rathakrishnan.
© 2021 John Wiley & Sons, Inc. Published 2021 by John Wiley & Sons, Inc.
Companion Website: www.wiley.com/go/Rathakrishnan/IntroductiontoAerospaceEngineering

Table A.1 (Continued)

	SI	US
Force/weight	Newton (N)	Pound (lb) = 4.448 N
Pressure	Pascal (Pa) = 1 N/m^2	
	Kilopascal (kPa) = 10^3 Pa	Pounds per square inch (psi) = 6.893 kPa
	Torricelli (torr) = 1 mm of	Inches of mercury (in. Hg) =
	Mercury (mm Hg) = 0.1333 kPa	25.4 mm Hg = 0.491 psi
Energy	Joule (J)	Foot-pound (ft-lb) = 1.356 J
	Calorie (cal) = 4.187 J	
	Kilocalorie (kcal or Cal) = 10^3 cal	British thermal unit (Btu) = 1055
		J = 0.252 kcal
Power	Watt (W) = joules per second (J/s)	
	Kilowatt (kW) = 10^3 W	horsepower (hp) = 0.7457 kW
Moment of force or torque	N-m	Pound-foot ((lb-ft) = 1.3558 N-m)

Index

Introduction to Aerospace Engineering: Basic Principles of Flight, First Edition. Ethirajan Rathakrishnan.
© 2021 John Wiley & Sons, Inc. Published 2021 by John Wiley & Sons, Inc.
Companion Website: www.wiley.com/go/Rathakrishnan/IntroductiontoAerospaceEngineering